SCIENCE REFERENCE GUIDES

Di Barton
Mike Evans

SAINT BENEDICT SCHOOL
DUFFIELD ROAD
DERBY DE22 1JD

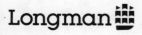

Longman Group UK Limited,
*Longman House, Burnt Mill, Harlow,
Essex CM20 2JE, England
and Associated Companies throughout the world.*

© Longman Group UK Limited 1989
*All rights reserved; no part of this publication may be
reproduced stored in a retrieval system, or transmitted
in any form or by any means, electronic, mechanical,
photocopying, recording, or otherwise without either
the prior written permission of the Publishers or a licence
permitting restricted copying in the United Kingdom
issued by the Copyright Licensing Agency Ltd,
90 Tottenham Court Road, London W1P 9HE.*

First published 1989
Fourth Impression 1994

British Library Catologuing in Publication Data

Barton, Di
   Science.
   1. Science
   I. Title . Evans, Michael
   500

   ISBN 0-582-05076-6

Designed and produced by The Pen and Ink Book Company Ltd, Huntingdon, Cambridgeshire

Illustrated by Chris Etheridge

Set in 9/10pt Century Old Style

Printed and bound in Singapore

HOW TO USE THIS BOOK

Throughout your GCSE course you will be coming across terms, ideas and definitions that are unfamiliar to you. The Longman Reference Guides provide a quick, easy-to-use source of information, fact and opinion. Each main term is listed alphabetically and, where appropriate, cross-referenced to related terms.

- Where a term or phrase appears in **different type** you can look up a separate entry under that heading elsewhere in the book.
- Where a term or phrase appears in **different type** and is set between two arrowhead symbols ◄ ►, it is particularly recommended that you turn to the entry for that heading.

ABIOTIC FACTORS

Abiotic factors are the non-living or physical factors which affect the organisms in an **ecosystem**, such as temperature, rainfall, light intensity, wind, the type of soil and the **nutrient cycles** of **nitrogen**, **carbon** and **water**.

If you have studied ecology at school you will know that the abiotic factors can affect where plants and animals live. For example, the amount of light and the type of soil can affect the *growth* of plants. The amount of exposure on rocky shores can affect the *distribution* of different animals and plants.

◀ Carbon, ecosystem, nitrogen, nutrient cycles, water ▶

ABSOLUTE ZERO

◀ Kelvin, gas laws ▶

ABSORPTION

Absorption of food takes place in the small intestine (ileum). The small molecules of **glucose** and **amino acids** move through the wall of the gut into the blood stream. The surface of the small intestine has been increased in two ways. First, it is very long, and second, it is covered with tiny finger-like villi. Both of these features *increase* the surface area so that more molecules can be absorbed through the thin wall of the gut and into the blood stream (Fig A.1). ◀ Amino acids, glucose ▶

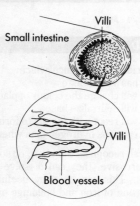

Fig A.1 Absorption; the villi in the small intestine increase the area for absorption

ACCELERATION

ACCELERATION

Acceleration is the increase in **velocity** per unit time. To find acceleration divide the increase in velocity by the time taken to increase it. For example, if a car increases its velocity from 20m/s to 40m/s in 10 seconds, then its acceleration is $\frac{(40 - 20)}{10} = 2\text{m/s}^2$

The units of acceleration are metres per second per second (Fig A.2).
◀ Velocity ▶

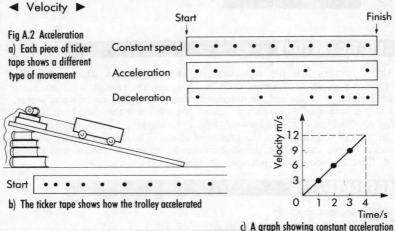

Fig A.2 Acceleration
a) Each piece of ticker tape shows a different type of movement

b) The ticker tape shows how the trolley accelerated

c) A graph showing constant acceleration

ACCOMMODATION

◀ Eye ▶

ACID

An acid is a substance which:
- has a sour taste
- will change the colour of plant dyes (indicators)
- will neutralize bases
- will react with metals to form salts
- contains hydrogen ions when dissolved in water

We can recognize the sharp or sour taste in fruits or in vinegar. Acids can also be corrosive, they can dissolve many substances such as metals and some rocks. This can be very useful but can also be a nuisance, as with the corrosion of iron and buildings by acids in the atmosphere, e.g. **acid rain**.

 ACIDS AND WATER

Acids in water contain hydrogen ions

Acids only behave as acids when they are dissolved in water. So we only really meet them as solutions in water. The reason for this is that, in water, the acid produces hydrogen ions (H^+). All acids contain hydrogen ions in water. The importance of water can be demonstrated by dissolving some citric acid crystals (the acid from fruits such as oranges and lemons) in water and some citric acid crystals in another solvent such as *propanone* (nail varnish remover). When the citric acid crystals are dissolved in propanone the solution will *not* affect the colour of indicators nor will it react with bases or metals. However, in *water* the citric acid behaves as an acid, changing the colour of the indicator.

In fact the acid reacts with the water to produce a hydroxonium ion (H_3O^+):

$$H^+ + H_2O \rightarrow H_3O^+$$

It is really this ion that is responsible for acidity; however, in order to keep things simple we can think of acids as just providing hydrogen ions in water. The following ions can be regarded as being equivalent: H^+, $H^+(aq)$, $H_3O^+(aq)$.

Strong acids and weak acids

Those acids which provide a lot of hydrogen ions in water are called *strong* acids, whereas those which only provide a few are called *weak* acids. Examples of strong acids and weak acids are shown in Table A.1.

Strong acids	*Weak acids*
sulphuric acid	citric acid (citrus fruits)
nitric acid	ethanoic acid (acetic) (vinegar)
hydrochloric acid	malic acid (apples)

Table A.1 Strong acids and weak acids

Ethanoic acid (acetic acid) is a **weak** acid because only a proportion of its molecules in solution split up to provide hydrogen ions.

Some common acids

Some common acids are shown in Table A.2.

Acid	*Formula*	*Ions present*
hydrochloric acid	HCl	H^+ Cl^- (chloride)
sulphuric acid	H_2SO_4	$2H^+$ SO_4^{2-} (sulphate)
nitric acid	HNO_3	H^+ NO_3^- (nitrate)
ethanoic acid	CH_3COOH	H^+ CH_3COO^- (ethanoate)

Table A.2 Common acids

Notice that the names of acids are taken from the **anion** present. This ion is referred to as the *acid radical*.

Concentration of acid solutions

Acid solutions can be *concentrated* or *dilute* depending upon how much water is present. The concentration of a solution can be measured in *molarity* mol/dm^3 or g/dm^3. For example a 1M (1 molar) solution of sulphuric acid contains 98g H_2SO_4 per dm^3.

4 ACID

This is calculated from the formula mass of $H_2SO_4 = (2 \times 1) + 32 + (4 \times 16) = 98$. Do not confuse *strong* acids with *concentrated* acids. We can have concentrated solutions of strong and weak acids, as well as dilute solutions of strong and weak acids. The acids that are used in a laboratory are dilute acids of concentrations 2M, 1M, 0.1M etc.

ACIDS AND INDICATORS

Acids change the colour of some dyes which we call **indicators**. Acids will turn litmus indicator red and Universal Indicator red, orange or yellow. Strong acids have a low pH number and will turn Universal Indicator red, whereas weak acids turn Universal Indicator yellow.

PATTERNS OF ACID REACTIONS

Acids are very useful substances because they react with a large number of other substances in fairly predictable ways. They are used extensively in industry in the manufacture of a large variety of materials.

Reactions with metals

General pattern of reaction:

| Acid | + | metal | → | salt | + | hydrogen |

Examples:

| zinc | + | hydrochloric acid | → | zinc chloride | + | hydrogen |
| Zn | + | 2HCl | → | $ZnCl_2$ | + | H_2 |

| magnesium | + | sulphuric acid | → | magnesium sulphate | + | hydrogen |
| Mg | + | H_2SO_4 | → | $MgSO_4$ | + | H_2 |

The solution that is produced is *neutral* and the salt produced depends on the acids. Some **metals** react with acids faster than others. The metal's reactivity depends on its position in the **reactivity series**:

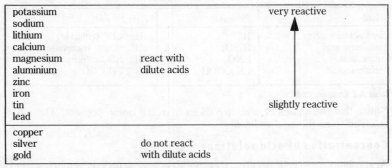

Note: The gas hydrogen can be tested for; if it is present in a test tube, a lighted splint will produce a 'pop'.

Reactions with metal oxides

Metal oxides are **bases** and react with acids to produce salt and water. All metal oxides will react with dilute acids.
General pattern of reaction:

| metal oxide | + | acid | → | salt | + | water |

Examples:

copper (II) oxide	+	nitric acid	→	copper nitrate	+	water
CuO	+	$2HNO_3$	→	$Cu(NO_3)_2$	+	H_2O
magnesium oxide	+	sulphuric acid	→	magnesium sulphate	+	water
MgO	+	H_2SO_4	→	$MgSO_4$	+	H_2O

Reactions with metal hydroxides

Metal hydroxides are bases which will react with acids to produce salt and water. Metal hydroxides which are soluble are called **alkalis**. All metal hydroxides react with acids.
General pattern of reaction:

| metal hydroxide | + | acid | → | salt | + | water |
| $OH^-(aq)$ | + | $H^+(aq)$ | → | $H_2O(l)$ | | |

Examples:

calcium hydroxide	+	hydrochloric acid	→	calcium chloride	+	water
$Ca(OH)_2$	+	$2HCl$	→	$CaCl_2$	+	H_2O
potassium hydroxide	+	sulphuric acid	→	potassium sulphate	+	water
$2KOH$	+	H_2SO_4	→	K_2SO_4	+	$2H_2O$

Reaction with carbonates

Metal carbonates can be thought of as bases. They too will react with acids to produce salt and water, but they also produce **carbon dioxide**.
General pattern of reaction:

| acid | + | carbonate | → | salt | + | water | + | carbon dioxide |

Examples:

sodium carbonate	+	nitric acid	→	sodium nitrate	+	water	+	carbon dioxide
Na_2CO_3	+	$2HNO_3$	→	$2NaNO_3$	+	H_2O	+	CO_2
calcium carbonate	+	hydrochloric acid	→	calcium chloride	+	water	+	carbon dioxide
$CaCO_3$	+	$2HCl$	→	$CaCl_2$	+	H_2O	+	CO_2

ACID

This is a general pattern for all carbonates. Since many rocks are carbonates, it can be used as a test to help identify rocks. For example, limestone, marble and chalk are all mainly calcium carbonate and will therefore react with hydrochloric acid. When a small amount of acid is placed on the surface, the rocks 'fizz' and give off carbon dioxide.

Note: There is an exception; sulphuric acid will *not* react well with calcium carbonate rock since, during the reaction, a layer of calcium sulphate builds up on the surface, which being insoluble in sulphuric acid, prevents any further reaction. Carbon dioxide can be tested for. When the gas is bubbled through limewater, the limewater turns cloudy.

ACIDS IN ACTION

Acids are corrosive and can be a nuisance; however, they can also be a benefit. For example, acids are used in preserving food (in pickling and chutneys), and are present in the digestive system to assist with the breakdown of food.

Acids and food preservation

The 'pickling' of foods, e.g. onions and eggs, and the making of chutneys helps to preserve the food. The acid that is used is found in vinegar. This acid is called *ethanoic acid* or *acetic acid* (acetic acid is the common name). The reason it works is that any bacteria which enter the food are killed by dehydration due to **osmosis**. It also means that the **pH** is too low for **enzymes** (biological catalysts) to work, thereby preventing the natural deterioration of food. Also, adding lemon juice to sliced apples prevents them from going brown, since the acidity of the lemon juice stops the enzymes in the apple from working.

Acids and digestion

The stomach lining produces gastric juice. This contains *hydrochloric acid* (about pH 2) which kills most of the micro-organisms in the food. Gastric juice also contains the enzyme *pepsin*, which starts to break down large protein molecules. Pepsin obviously works best at a much lower pH than salivary amylase.

Indigestion is often caused by too much acid in the stomach and can be relieved by taking 'antacid' tablets. These contain bases which neutralize the excess acid in the stomach. Examples are Settlers and Rennies, which both contain calcium carbonate and magnesium carbonate.

ACIDITY AND THE SOIL

Most plants prefer to grow in a soil which is slightly acidic, about pH 6 – pH 7. There are even some plants which prefer more acidic soils (pH 4.5 – pH 6) such as azaleas, rhododendrons and heathers. No plants will grow in strongly alkaline soils, however, although some plants will grow in weakly alkaline soils (up to pH 8).

ACID RAIN

When changing from growing one type of plant to another, sometimes the acidity of the soil has to be changed to get the best results. In addition, the soil acidity itself may well change over a period of time due to the plants themselves. Needing to reduce the acidity of the soil is a common problem to farmers. They overcome this by adding 'lime' to the soil. Lime is calcium oxide, though calcium hydroxide (slaked lime) or calcium carbonate (limestone or chalk) is often used. The reactions involved are as follows:

Lime or quicklime
$$CaO + 2H^+ \rightarrow Ca^{2+} + H_2O$$

Slaked lime
$$Ca(OH)_2 + 2H^+ \rightarrow Ca^{2+} + 2H_2O$$

Limestone
$$CaCO_3 + 2H^+ \rightarrow Ca^{2+} + H_2O + CO_2$$

◀ Acid rain, digestion ▶

ACID RAIN

Normally, rain is very slightly acidic (pH 5), due to a small amount of carbon dioxide from the atmosphere which dissolves in the rain water to produce a weakly acidic solution.

$$H_2O + CO_2 \rightleftharpoons H_2CO_3$$

Acid rain, however, has a pH of between 5 and about 2.2. The strongest acid rain has an acidity comparable to that of lemon juice. The causes of acid rain are not fully understood, but enough is known to realise that the burning of **fossil fuels** (hydrocarbons) and the exhaust emissions from cars contribute greatly to acid rain. Fossil fuels, like coal and oil, contain impurities of sulphur, so that when they burn they produce *sulphur dioxide* in addition to the normal products of combustion (carbon dioxide and water). Sulphur dioxide is an acidic gas.

The exhausts of cars also emit gases other than the normal products of combustion. In the car engine where the petrol (hydrocarbon fuel) is burned, the temperature is so high that nitrogen from the air reacts with oxygen to form *oxides of nitrogen* which escape through the exhaust, together with unburned hydrocarbons and carbon monoxide. This mixture of gases (particularly sulphur dioxide and the oxides of nitrogen) may react in the atmosphere with ozone to produce rain which contains sulphuric and nitric acids. It is these substances which give the rain its acidity.

Acid rain is a greater problem in parts of the world which are industrialized and therefore burn fossil fuels in power stations and factories and have large numbers of cars, e.g. Europe, USA, Canada, USSR.

EFFECTS OF ACID RAIN

Acid rain will corrode metals and will react with some building materials (limestone and marble), gradually eating them away. Perhaps the largest worry, however, is the effect it has either directly, or indirectly, on living things. When acid rain falls onto the soil it dissolves away many of the minerals (salts) in the soil. These contain *metal ions*, which are 'leached' (dissolved) out of the soil and washed into rivers and lakes. These then become increasingly acidic due to the rising concentration of metal ions. The first metal ions to be removed from the soil are those that dissolve more easily, magnesium (Mg^{2+}) and calcium (Ca^{2+}). These ions are needed by plants to ensure healthy growth, so are no longer available to the plants.

Metal ions such as aluminium (Al^{3+}), lead (Pb^{2+}) and copper (Cu^{2+}) are the next to be leached out of the soil. These ions are particularly troublesome since they are poisonous. Aluminium, when dissolved in the water, prevents the gills of fish working, as well as being poisonous to other organisms. (*Note*: it is not the acid water that kills the fish but the dissolved metal ions).

The areas that are affected more than others are those which lie on *thin soil* and *granite rock*, e.g. Scotland, Dartmoor, the Black Forest in Germany and Scandinavia. Areas which are lucky enough to have deep soil covering limestone rock are not so badly affected, since the limestone rock and its soil can neutralize the effects of the acid rain. How an area is affected depends on its position, since the acidic gases are carried on prevailing winds.

COMBATTING ACID RAIN

1. Some lakes which are very acidic have large amounts of lime (calcium hydroxide) added to them to neutralize the acidity. This is only a temporary measure, however, since it has to be frequently repeated and costs a large amount of money.
2. Power stations which burn coal and oil can be fitted with equipment which will *remove* the sulphur dioxide from the gases before they are released into the atmosphere.
3. Car exhausts can be fitted with 'catalytic converters' which can convert the *harmful* gases into *harmless* ones. These cannot be fitted to the majority of cars in this country, however, since most cars run on leaded petrol. The lead in the petrol will prevent the catalyst from working.

ACTIVATION ENERGY

Almost all chemical reactions need a 'push', in other words an amount of energy to get them going. This *initial* amount of energy can be quite small or quite large, and is called the *activation energy*. It is the energy required to initially break the chemical bonds in order to allow a reaction to proceed. Fuels such as gas and coal need to be supplied with a source of heat (from a match) to start the combustion. This activation energy can be represented on energy level diagrams (Fig A.3).

◀ Endothermic; Exothermic ▶

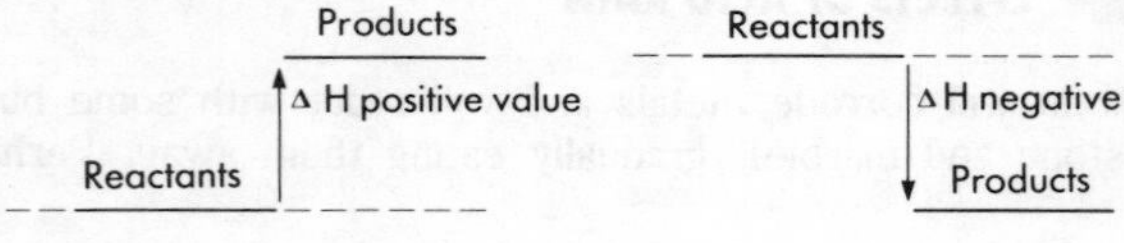

Fig A.3 Activation energy

Reactants → Products, ΔH positive value

Reactants → Products, ΔH negative value

ACTUATOR

An *actuator* brings about the correction of a **feedback system**. It is the effector device which carries out a change. For example, in an oven the actuator is the heating element. When the heating element gets hot, it brings the temperature up to the set point. ◀ **Feedback system** ▶

ADRENALIN

Adrenalin is a **hormone** which is secreted by the adrenal glands in response to demands made on your body. Adrenalin has several effects on the body. These include: rapid pulse, deep, rapid breathing, blood being diverted from the skin to the muscles, and stores of carbohydrates in the liver being converted from glycogen to **glucose** to prepare the body for action, such as running away from a frightening situation. ◀ **Glucose, hormone** ▶

AEROBIC RESPIRATION

Aerobic respiration is the breakdown of carbohydrates and fats to release energy, using oxygen. The word equation below summarises this process:

food + oxygen → carbon dioxide + water + energy

The chemical equation for this process, with *glucose* as food, is:

$$C_6H_{12}O_6 + 6O_2 \rightarrow 6CO_2 + 6H_2O + 2830 \text{ kJ}$$

Aerobic respiration takes place in the cells in your body, for example the muscle cells. Your blood carries food and oxygen to the cells, and transports the waste products, carbon dioxide and water, from the cells. These waste products are removed by the lungs when you breathe out.

One way of investigating aerobic respiration in living organisms is to identify the carbon dioxide produced. For example, the gas produced by a mouse can be bubbled through limewater. If the limewater turns cloudy, then the gas is carbon dioxide (Fig A.4). ◀ **Breathing** ▶

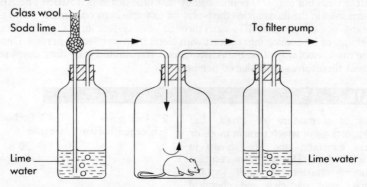

Fig A.4 Testing for aerobic respiration

AEROSOL

A can of *aerosol* contains a product (deodorant, etc.) and a liquefied propellant gas under pressure. When the nozzle is pressed, the pressure in the can forces out a fine spray of the propellant mixed with the product. The propellant, no longer under pressure, very quickly evaporates, leaving the product (deodorant, polish, glue, etc.). There are two main types of propellant, hydrocarbons and chlorofluorocarbons.

- *Hydrocarbons* (usually propane or butane) are often used in furniture polish and air fresheners
- *Chlorofluorocarbons* (CFCs) are more often found in personal care products (deodorants, hairspray, etc.). These are at present being phased out because they damage the **ozone layer.**

AEROSOL SAFETY

The aerosol contents are under pressure so there is a risk of an explosion if it is overheated or punctured. In addition, hydrocarbons are very flammable; this is an important reason why CFCs started to be used as propellants because they were not flammable, but these of course have other problems (Fig A.5). ◄ Ozone layer ►

Fig A.5 Aerosol

AIDS (ACQUIRED IMMUNE DEFICIENCY SYNDROME)

AIDS is caused by a virus which attacks the body's defence system that normally helps you to fight infection. In 1986 40,000 people in Britain had the AIDS virus but only 600 people had by that time developed AIDS. The AIDS virus lives in the fluids inside the body; for example in blood, semen, saliva and tears. You can get the AIDS virus through sexual intercourse with an infected person, or by coming into contact with blood from an infected person. One of the most usual ways that the AIDS virus is transmitted is by drug users who inject themselves with shared needles.

AIR

Air is a mixture of gases, the proportions of which remain more or less constant; and are shown in Figure A.6. The air also contains varying amounts of water vapour, dust, soot particles and chemical pollutants such as sulphur dioxide.

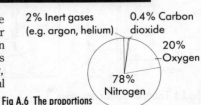

Fig A.6 The proportions of gases in air

AIR PRESSURE

Animals and plants are continually **respiring**, that is, taking in oxygen and using it to convert food into energy; the waste product, carbon dioxide, is released into the atmosphere. During the process of photosynthesis green plants take in carbon dioxide and combine it with water using the sun's energy to produce simple sugars. The waste product is oxygen. Whilst these processes are in balance, the proportion of gases will remain the same; however, Man's burning of **fossil fuels** (in power stations, cars etc.) releases more carbon dioxide into the atmosphere as well as other gases. In addition Man is cutting down large areas of tropical rain forest and hence reducing the global potential for **photosynthesis**.

AIR PRESSURE

The Earth's gravitational pull holds the layer of gases around the surface of the planet, and the weight of air above any part of the Earth is described as air pressure. For example, the weight of air pressing down on 1 square centimetre is 10 **newtons** (N). Atmospheric pressure is therefore about 10 N/cm^2. This unit is also described as one bar or 1000 millibars.

Pressure is usually measured with a *barometer*, but in an aeroplane a pilot uses a sensitive barometer, called an *altimeter*, which measures atmospheric pressure and height. Atmospheric pressure decreases with altitude. For example, as you go higher up a mountain there is less air above you pressing down on you. If you were to go about 6 kilometres high, the pressure drops to about half that at sea level as the air molecules thin out.

▶ AIR PRESSURE AND PARTICLES

The *particles* in a gas such as air are moving very fast. When these particles hit something they exert a **force** on that object. The combined effect of the many millions of particles in a gas acting on a particular area is its pressure.
Pressure = force per unit area.
(Units of pressure are Newtons per square metre; N/m^2)

The *pressure of a gas can be increased* by increasing its temperature (heating). The gas particles will have more **kinetic energy** and so will be moving faster and striking the sides of a container harder and more often.

The *pressure of a gas can be increased* by *reducing the volume of that gas*. The particles in the gas will be closer together so they will strike the walls of the container more often.

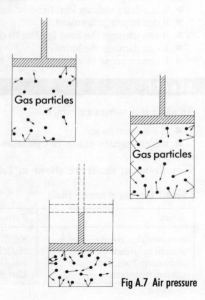

Fig A.7 Air pressure

Remember: the *volume of a gas can be increased* by *increasing the temperature* (expansion)

The following ideas can help explain the '**gas laws**' or 'gas patterns' (α means proportional):
1. Pressure α Temperature (provided the volume remains the same)
2. Pressure α 1/Volume (provided the temperature remains the same)

These can be re-expressed as: P/T = constant, and
P$\times$V = constant.

In order to make calculations using these patterns or laws, temperature is measured on the **Kelvin** scale of absolute temperature. 0 K is known as absolute zero and is the temperature at which particles have no kinetic energy. 0 K = $-273°$ Celsius.

ALCOHOL

The *alcohols* are a group of carbon compounds which contain a hydroxyl group (OH). Some examples of alcohols are shown below:

Name	Formula
Methanol	CH_3OH
Ethanol	C_2H_5OH
Propanol	C_3H_7OH

Ethanol is the alcohol we find in drinks (beer, wine, whisky, etc.). It is produced by the fermentation of sugars using **yeast**. Ethanol (alcohol) can affect the body in a number of ways:
- It is a drug and can lead to addiction.
- It can impair judgement.
- It can damage the liver leading to cirrhosis.
- It can damage the kidneys.
- It can increase blood pressure.

ALKALI

An alkali is a substance which:
- is a soluble base
- will change the colour of indicators
- will neutralize acids
- contains hydroxide ions.

Some common alkalis are shown in Table A.3.

Table A.3 Examples of common alkalis

Alkali	Formula	Ions present		
potassium hydroxide	KOH	K^+	(potassium)	OH^-
sodium hydroxide	NaOH	Na^+	(sodium)	OH^-
ammonium hydroxide	NH_4OH	NH_4^+	(ammonium)	OH^-
calcium hydroxide	$Ca(OH)_2$	Ca^{2+}	(calcium)	$2OH^-$

ALKALI METALS

In the same way that there are strong and weak *acids* so there are strong and weak *alkalis*. Strong alkalis provide a lot of free hydroxide ions in solution. Sodium and potassium hydroxides are *strong alkalis*, whereas an aqueous solution of ammonia (ammonium hydroxide) and calcium hydroxide are *weak alkalis*.

Alkalis turn litmus indicator blue, and will turn universal indicator blue or violet depending on their strength; violet indicates strong alkalis with a high pH number. ◀ pH scale ▶

Alkalis will also neutralise acids:

$$H^+ + OH^- \rightarrow H_2O$$

Example:

Hydrochloric acid + sodium hydroxide → sodium chloride + water

HCl + NaOH → NaCl + H_2O

ALKALI METALS

Each group in the **periodic table** contains elements which behave in similar ways in chemical reactions. This is because they have the **same number of electrons in their outer shells**. They do, however, differ by degrees in their intensity of reaction as you travel 'down the group' (Fig A.8).

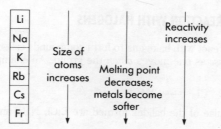

Fig A.8 The alkali metals

Group 1 elements are known as the *alkali metals*. These appear in the first column of the periodic table.

ELECTRON STRUCTURE

All elements in Group 1 have one electron in the outer shell. Their electron structures are as follows:

Lithium	Li	2,1
Sodium	Na	2,8,1
Potassium	K	2,8,8,1
Rubidium	Rb	2,8,18,8,1
Caesium	Cs	2,8,18,18,8,1
Francium	Fr	2,8,18,18,18,8,1

ALKALINE EARTH METALS

REACTION WITH AIR

Lithium, sodium and potassium react with air to form oxides. When the metal is cut with a knife, its surface quickly tarnishes. The *reactivity* (speed of reaction) *increases* as you move *down* the group. The general reaction between alkali metals and oxygen is as follows:

$$4M + O_2 \rightarrow 2M_2O$$

The formulae of the oxides formed are Li_2O, Na_2O and K_2O. These dissolve in water to produce alkaline solutions.

REACTION WITH WATER

Lithium, sodium and potassium react quickly with water. Lithium when placed on water 'fizzes' and quickly reacts. Sodium will buzz around on the surface of the water, giving the odd spark. Potassium reacts more violently, burning with a lilac flame. *Reactivity increases* as you move *down* the group. Each produces a strong alkaline solution with water. The general reaction is as follows:

$$2M + 2H_2O \rightarrow 2MOH + H_2$$

The formulae of the hydroxides produced are LiOH, NaOH and KOH.

REACTION WITH HALOGENS

Each will react with halogens to form compounds called halides. The *reactivity will increase* as one moves *down* the group. An example is the reaction with chlorine:

$$2M + Cl_2 \rightarrow 2MCl$$

The formulae of the halides formed are LiCl, NaCl and KCl.

REACTIVITY TRENDS

The atoms react to form ions which have a charge of 1+. In forming these ions, for example Na^+, an electron has to be removed from the outer shell. Those atoms which have outer shells further away from the positive nucleus (the bigger atoms) will require less energy to remove that electron, so will tend to be more reactive; hence *reactivity increases* as you move *down* the group: Fr>Cs>Rb>K>Na>Li.

ALKALINE EARTH METALS

Group 2 elements are known as the *alkaline earth metals*. These appear in the second column of the periodic table.

ALKALINE EARTH METALS

 ELECTRON STRUCTURE

All elements in Group 2 have two electrons in the outer shell. Their electron structures are as follows:

Beryllium	Be	2,2
Magnesium	Mg	2,8,2
Calcium	Ca	2,8,8,2
Strontium	Sr	2,8,18,8,2
Barium	Ba	2,8,18,18,8,2
Radium	Ra	2,8,18,18,18,8,2

 REACTION WITH AIR

The *reactivity increases* as you move *down* the group. The general reaction is as follows:

$$2M + O_2 \rightarrow 2MO$$

The formulae of the oxides produced are CaO and MgO. These oxides are bases.

 REACTION WITH WATER

The *reactivity increases* as you move *down* the group. Magnesium will not react with cold water, but reacts with steam:

$$Mg(s) + H_2O(g) \rightarrow MgO(s) + H_2(g)$$

Calcium will react with cold water:

$$Ca(s) + 2H_2O(l) \rightarrow Ca(OH)_2(aq) + H_2(g)$$

 REACTION WITH DILUTE ACIDS

The metals will react with dilute acids, *reactivity increases* as you move *down* the group. For example:

$$Mg + 2HCl \rightarrow MgCl_2 + H_2$$

 REACTIVITY TRENDS

The atoms react to form ions which have a charge of 2+. In forming these ions, for example Ca^{2+}, two electrons have to be removed from the outer shell. Those atoms which have outer shells further away from the positive nucleus (the bigger atoms) will require less energy to remove that electron so will tend to be more reactive; hence reactivity increases as you move down the group. Also since two electrons have to be removed rather than one to form ions, these metals are *less* reactive than the *alkali metals* in Group 1.

ALLOTROPE

Some elements can exist in different forms or allotropes, e.g. giving rise to different shape crystals, depending on how their atoms can pack together. Sulphur has two allotropic forms (Fig A.9a)). *Rhombic* sulphur is a crystalline form which is stable at room temperature, whereas *monoclinic* sulphur is stable at 96°C (and above, before melting). Carbon can exist as the two allotropic forms *graphite* and *diamond* (Fig A.9b)).

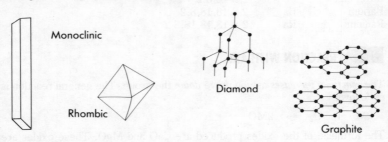

Fig A.9 a) Allotropes of sulphur   b) Allotropes of carbon

Allotropes can be *converted* from one form into the other, e.g. graphite could be converted to diamond (one of the hardest substances known) by heating under extreme pressure (about 15,000 atmospheres at 300°C).

ALLOYS

Alloys are solid mixtures of metals and are formed by melting together two, or more, different metals. The properties of the alloy that is formed are *not* simply an average of the properties of the original metals. The properties also depend on the proportion of each metal present; for example, *solder* is an alloy of lead (67%) and zinc (33%). It has a lower melting point and is softer than either lead or zinc. This is not to say, however, that the properties of 'new' alloys are simply a hit or miss affair. The properties of alloys are determined partly by the metal atoms present, but more importantly by how these atoms can pack together in the structure. With information such as this, scientists can produce alloys with *specific properties*.

One very important group of alloys are **steels**. Steel is an alloy of the metal *iron* mixed with a small quantity of *carbon* (a non-metal). The introduction of a small quantity of carbon increases the strength of the iron enormously because the carbon atoms fit into the giant structure of iron atoms, preventing the iron atoms moving so freely when the material is hammered, twisted or stretched (Fig A.10).

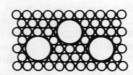

Fig A.10 Alloys

Different steels can be produced, each with different properties, depending on the amount of carbon present. In addition other metals can be added, such as chromium and tungsten, to change the properties still further. Some examples of common alloys are shown in Table A.4.

Table A.4 Examples of common alloys

Alloy	Use	Constituents	Important properties
spring steel	suspension springs	iron; 0.3%-0.6% carbon	contains sufficient carbon that will produce a springy metal
stainless steel	surgical instruments; cutlery	iron < 1% carbon 18% chromium	resistant to corrosion
chromium-vanadium steel	axles and wrenches	iron; chromium; vanadium; carbon	very strong, great resistance to strain
high speed tungsten steels	cutting metals; drills etc	iron with up to 20% tungsten	maintains sharp edge at high temperatures
brass	screws; taps; ornaments	copper/zinc. The more zinc there is, the stronger the alloy, up to 34% zinc	strong; does not easily corrode
bronze	castings of intricate shapes, statues, etc.	copper up to 12% tin	easily cast; resists corrosion
an aluminium alloy	aircraft framework	aluminium; copper; magnesium	very strong for its weight; aluminium is a light metal but is not very strong, so is alloyed with other metals to increase strength

ALPHA RADIATION (or ALPHA PARTICLES)

When the nucleus in radioactive materials breaks down it can emit different types of radiation. This radiation is of three types, alpha, **beta** and **gamma**. Each type of radiation has characteristically different properties (Fig A.11).

Alpha (α) particles are fast moving helium nuclei (groups of two protons and two neutrons). They are easily stopped by a sheet of paper and will not travel very far through the air. They are weakly deflected by a magnetic field. ◄ Beta radiation, Gamma radiation ►

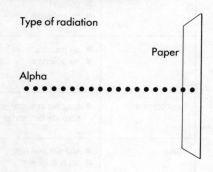

Fig A.11 Alpha radiation

ALTERNATING CURRENT (AC)

Alternating current describes an electric current which changes direction and is produced by a **generator**. The current increases to a certain value in one direction; it then decreases and reaches the same value in the opposite direction (Fig A.12). The *frequency* of the current is the number of complete changes made in one second. For example, mains electricity is generated at a frequency of 50 cycles a second, or 50 **hertz**.

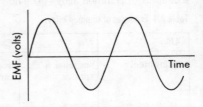

Fig A.12   Alternating current

ALTERNATIVE ENERGY

This term is used to describe any source of energy which does not involve using **fossil fuels** such as coal, oil or gas. For example, **solar energy** is obtained from the Sun; **hydro-electric power** is obtained from fast-flowing rivers; **tidal energy** and **wave energy** are obtained from the movement of water; **wind energy** is obtained from the wind, and **geothermal energy** comes from the heat deep inside the Earth.

Table A.5 Alternative energy

Energy Source	Advantage	Disadvantage
Wind	<ul><li>will not run out</li><li>no fuel costs</li><li>no pollution</li><li>useful for isolated communities</li></ul>	<ul><li>windmills can spoil the environment</li><li>wind speeds may vary, so generation of electricity is varied</li></ul>
Solar	<ul><li>will not run out</li><li>no costs</li><li>no pollution</li></ul>	<ul><li>cloud cover blocks sun</li><li>difficult to store energy produced</li><li>huge solar panels needed</li></ul>
Tidal	<ul><li>no fuel costs</li><li>no pollution</li></ul>	<ul><li>expensive to build power stations</li><li>may cause silting up of rivers</li></ul>
Geothermal	<ul><li>long term supplies can provide hot water</li></ul>	<ul><li>not easily available</li><li>costly to obtain</li></ul>
Wave	<ul><li>will not run out</li><li>no pollution</li></ul>	<ul><li>many technological problems</li><li>hazard to shipping</li></ul>

AMINO ACID

Table A.5 summarises the main alternative sources of energy, and their advantages and disadvantages.

There are **three** main reasons why alternative energy sources are being developed:

1. Fossil fuels are 'finite', which means that they will run out and cannot be replaced. It has been predicted that the coal supplies which are known about will have been used up in about 600 years.
2. Fossil fuels cause pollution as they produce sulphur dioxide, nitrogen oxides and carbon dioxide when they burn. Sulphur dioxide is one of the main causes of 'acid rain'. Carbon dioxide is one of the main factors in the 'greenhouse effect'.
3. There is an increasing demand, both from industry and from consumers, for electricity. This demand may exceed the amount of electricity which can be supplied by existing power stations which burn fossil fuels.

ALTERNATOR

An alternator is a **generator** which produces **alternating current**. The alternating current is induced as a coil rotates between the permanent magnets. The coil is linked to the outside circuit by two carbon brushes which press against two carbon slip rings that are fixed to the end of the coil (Fig A.13). The current can be increased by four factors:

1. having more turns on the coil
2. using stronger magnets
3. winding the coil on a soft iron armature
4. rotating the coil at a higher speed

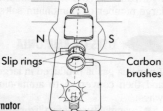

Fig A.13 Alternator

AMINO ACID

Amino acids are small molecules which are the building blocks of **proteins**. Proteins are made of long chains of amino acids. All amino acids contain the amino group $-NH_2$.

Fig A.14
Amino acids

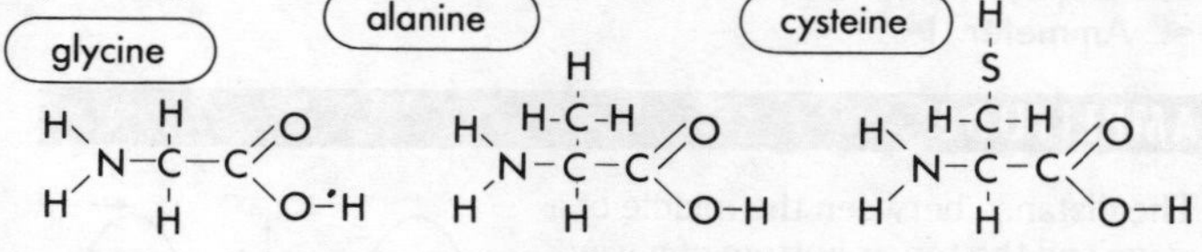

There are 22 different amino acids which can be combined to make thousands of different proteins. Three different amino acids are shown in Figure A.14. When protein is digested, the large molecules are broken down into amino acids which are then used by the body to form many different proteins, such as muscle, blood cells and hormones. ◀ **Protein** ▶

AMMETER

An ammeter is a device for measuring electric current in **amperes** (amps). It is always placed in series with a resistance or a circuit component through which the current flowing is to be measured. Ammeters should have a low **resistance** so that they do not reduce the flow of current.

AMMONIA

Ammonia's properties are as follows:
- it is a colourless gas
- it has a very strong, recognizable smell
- it is very soluble in water
- it is alkaline

Ammonia molecules each consist of one nitrogen and three hydrogen atoms; the formula is NH_3. Ammonia, as a gas, is difficult to handle, so it is more commonly found as a solution. When ammonia is passed through water it dissolves easily to form a weakly alkaline solution, called ammonium hydroxide.

ammonia + water → ammonium hydroxide
NH_3 + H_2O → NH_4OH

Ammonium hydroxide contains two ions – ammonium ion (NH_4^+) and hydroxide ion (OH^-). Ammonium hydroxide will react with acids to produce a large number of ammonium salts, many of which are very useful.

▶ USES OF AMMONIA

Ammonia gas is produced in large quantities every year by the *Haber Process* and then converted to ammonium compounds, etc. for use in a variety of ways:
- manufacture of fertilizers (approx. 80% is used for this)
- manufacture of nitric acid
- manufacture of some plastics
- manufacture of explosives
- manufacture of household cleaners

AMPERE

The ampere (amp) is the S.I. unit of electric current. The symbol used is A.
◀ Ammeter ▶

AMPLITUDE

The distance between the middle of a wave and the top or bottom of a wave is the amplitude. It is the amount by which a particle is displaced up and down. The distance marked A on Figure A.15 is the amplitude. ◀ Wavelength ▶

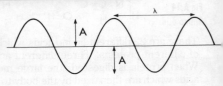

Fig A.15

AND GATE

AMYLASE

Amylase is a digestive enzyme produced by the salivary glands in the mouth and by the pancreas. Its function is to speed up the rate at which starch is broken down to *maltose* in the gut. Amylase works in an alkaline pH, so it doesn't work in the acid pH of the stomach. How amylase breaks down starch is shown in Figure A.16. Some textbooks refer to the amylase produced in the mouth as *ptyalin*. ◀ Digestion ▶

Fig A.16 How Amylase works

ANAEROBIC RESPIRATION

Anaerobic respiration is the breakdown of carbohydrates and fats to release energy, *without* using oxygen. The food is broken down to substances such as *lactic acid* and *alcohol*. Less energy is released compared to **aerobic respiration** which uses oxygen.

An example of anaerobic respiration occurs in your muscles when you are doing vigorous exercise. There is not enough oxygen supplied to your muscles to break down the food quickly enough and release energy needed by the body. Some energy is released from the food anaerobically and lactic acid is produced as a waste product. This is known as the 'oxygen debt'. When you stop the exercise, your rapid breathing provides extra oxygen to remove the lactic acid, 'repaying' the oxygen debt.

You can try producing lactic acid in your muscle by putting your arm up and quickly clenching and releasing the muscles of your hand. After a few 'fists' you may begin to feel pain in the muscles of your arm. This pain is caused by lactic acid. Athletes in sprint races usually use only anaerobic respiration to release energy quickly when they run, say, 100 metres.

Fig A.17

Some bacteria and fungi use anaerobic respiration and produce alcohol and carbon dioxide as waste products. This process is known as **fermentation**.

AND GATE

◀ Logic gates ▶

ANION

Substances which contain **ions** (charged particles) are electrically *neutral* (they contain equal numbers of positive and negative charges). Some examples of ionic compounds are M^+X^- or $N^{2+}Y^{2-}$.

Those ions which have a *negative* charge are called anions. Those which have a positive charge are called **cations**. Non-metal ions tend to be anions whereas metal ions tend to be cations.

During **electrolysis**, anions are attracted to the **anode** (positive electrode). Some examples of anions are shown below:

Name of ion	symbol	charge on ion
Oxide	O^{2-}	2–
chloride	Cl^-	1–
sulphide	S^{2-}	2–
sulphate	SO_4^{2-}	2–
carbonate	CO_3^{2-}	2–
nitrate	NO_3^-	1–
hydrogen carbonate	HCO_3^-	1–
phosphate	PO_4^{3-}	3–

ANODE

Anode is the name given to a *positive electrode* either in an **electrolysis** cell, or in an electrical cell (**battery**).

▶ THE ANODE IN ELECTROLYSIS

The *anode* and the cathode (the negative electrode) are attached to an electrical source (e.g. battery or power pack) and dipped into an *electrolyte* (liquid or solution containing ions). A current flows through the electrolyte when negative ions (**anions**) are attracted to the anode. At the anode these ions are converted to atoms:

$$X^- - e^- \rightarrow X$$

The anode must be able to conduct electricity and is usually carbon (graphite) or a metal. The anode that is chosen depends on the job the electrolysis cell has to do and the nature of the electrolyte; for example, often the anode must not react with the electrolyte. Sometimes, however, an anode is chosen that *will* 'react'. For example, impure copper can be purified if it is used as an anode in an electrolysis cell with copper sulphate as the electrolyte (Fig. A.18).

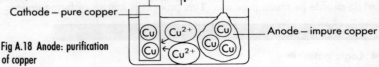

Fig A.18 Anode: purification of copper

The copper atoms at the anode lose electrons to form ions:

$Cu - 2e^- \rightarrow Cu^{2+}$

These ions are then transported through the electrolyte to the **cathode**. The cathode becomes coated with pure copper. The reaction occurring at the cathode is:

$Cu^{2+} + 2e^- \rightarrow Cu$

THE ANODE IN AN ELECTRICAL CELL

When two different metals (which act as electrodes) are placed in an electrolyte and are connected together externally, for example with a wire, they produce an electric current (a flow of electrons). The anode – the *positive* electrode – will be the metal which has the least tendency to lose electrons (it is lower in the reactivity series). In the *dry cells* (batteries) that we buy, the anode is not always a metal:

- In zinc-carbon batteries the anode is *carbon* (carbon is lower in the activity series than zinc).
- In 'alkali batteries' the anode is *manganese (IV) oxide*
- In 'calculator batteries' the anode is *mercury oxide*

ANODISING

This is the process of coating objects made of aluminium with a very thin layer of aluminium oxide. This layer protects the metal from corrosion but dulls the shiny surface of the aluminium. Anodising is carried out by **electrolysis** using sulphuric acid as the electrolyte which releases oxygen at the **anode**. The oxygen reacts with the surface of the aluminium, covering it with the oxide layer, which can be dyed to produce different coloured finishes.

ANTACID

Antacids (anti-acids) are products such as Rennies, Settlers etc., which are sold to relieve indigestion caused by too much acid in the stomach. Antacids neutralize the excess acid in the stomach and so relieve the indigestion. The active ingredient in antacids that can be bought is often calcium or magnesium carbonate. ◀ Acids, digestion ▶

ANTIBODY

These are chemical substances produced by **white blood cells**. Their function is to attack disease-causing bacteria in the body and to make the bacteria harmless. They are transported around the body by blood and lymph.

AORTA

The aorta is the main **artery** in your body and carries oxygenated blood away from the **heart** to the rest of your body.

AQUEOUS

An aqueous solution means one in which the solvent is *water*; for example, an aqueous solution of ammonia is produced by dissolving ammonia gas in water. In chemical equations, symbols of state can show whether a substance is dissolved in water. The state symbol for *aqueous* is aq. For example:

$$Mg(s) + 2HCl(aq) \rightarrow MgCl_2(aq) + H_2(g)$$

The hydrochloric acid (HCl) is dissolved in water (diluted). The magnesium chloride ($MgCl_2$) which is formed in the reaction is also dissolved in the water. This information is shown by the symbol (aq). ◄ State symbol ►

ARTERY

Arteries are blood vessels which carry blood *away* from the heart. The blood in the arteries is under pressure so the arteries have very thick elastic muscular walls (Fig A.19). The arteries divide into smaller blood vessels called *arterioles*, which then divide up to form tiny blood vessels called **capillaries**.

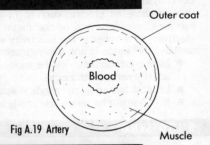

Fig A.19 Artery

ARTIFICIAL SELECTION

Artificial selection is how new varieties of animals and plants are produced by selective breeding. All the different varieties of dog have been produced by artificial selection of the wolf, a single wild species. Our domesticated cattle breeds, poultry, sheep and cereal crops have all been bred from wild species by generations of farmers. For example, a farmer may decide to cross-breed a sheep which produces high quality wool with a sheep which is very hardy and can tolerate harsh weather conditions. Similarly, horticulturists cross-pollinate plants to produce new varieties which are resistant to disease and have increased yields. Plant and animal breeding is now big business and so breeders are always on the look out for wild varieties that could be useful.

ASEXUAL REPRODUCTION

Asexual reproduction involves the production of a new generation of offspring from only one parent. This can be an advantage in that an isolated individual can reproduce on its own and produce offspring which are exact copies of itself. If the parent is successful then it is important for the survival of the offspring that characteristics are passed on exactly. However, a *lack* of **variation** can cause problems. For instance, if there is a sudden change in the environment, none of the offspring may survive; perhaps a disease might destroy the whole population because there would be no resistant varieties.

Bacteria, yeasts and other single-celled organisms reproduce asexually by growing to a maximum size and then dividing into two smaller individuals. A single disease-causing bacterium could do this every twenty minutes, so that in only twenty-four hours some 4000 million million bacteria would be produced, making you very ill indeed. Many different species of plants are able to reproduce asexually, but can also reproduce sexually depending on which method is most advantageous for survival of the species. Asexual reproduction in plants usually involves part of the plant becoming separated from the parent and developing into a new individual. Weeds in the garden do this when the gardener uses a mechanical cultivator, chopping up and replanting the weeds, thus accidentally increasing the weed problem. Gardeners are also able to grow new plants by taking cuttings, and yet be sure that the new plants will be just like the original (Fig A.20).

Examples of asexual reproduction include:
- a potato plant producing many new potatoes
- a strawberry plant producing a runner with many new stems growing up from it
- a cutting of a geranium which grows into a new plant.

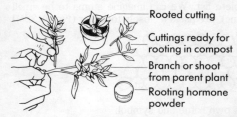

Fig A.20 Asexual reproduction in plants

ATMOSPHERE

The atmosphere is a layer of gases, a few hundred kilometres thick, which surrounds the Earth. The Earth is the only planet in the Solar System to have an atmosphere. The atmosphere surrounds the Earth and acts as a 'blanket' keeping the Earth warm. The Moon, which is the same distance from the Sun as the Earth, has no atmosphere. In the sunlight the Moon surface temperatures are as high as 100°C, but at night fall to −150°C. Our atmosphere is therefore very important to us in helping to keep the Earth's surface temperature more or less constant.

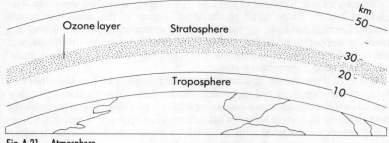

Fig A.21 Atmosphere

26 ATOM

The atmosphere is a mixture of gases (see **air**), and provides oxygen which is essential for **respiration**. It also absorbs harmful ultra-violet radiation from the Sun. The atmosphere has a series of different layers (Fig. A.21). Sunlight passes through the atmosphere to the Earth's surface which it warms. The warm surface of the Earth then re-radiates some of its heat into the lowest layer of the atmosphere, the *troposphere* (the layer where the weather occurs). The higher up you go, the cooler it becomes. At approx. 11 km the temperature is about −60°C. However, as you travel further up through the next layer, the *stratosphere*, the temperature starts to increase. This is because incoming sunlight is being absorbed in the **ozone layer** (within the stratosphere).

ATOM

All matter is made of three types of particles, atoms, **molecules** and **ions**. The atom is the most basic particle: there are just over 100 different atoms. The **periodic table** provides a complete list. We can imagine atoms to be small spheres which in a metal (e.g. zinc), are packed closely together (Fig. A.22a))

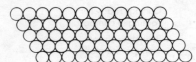

Fig A.22 a)

Each atom is represented by its own name and symbol:

Name	Symbol
Hydrogen	H
Oxygen	O
Carbon	C
Copper	Cu
Chlorine	Cl
Sodium	Na

Notice that each chemical symbol is either a single *capital* letter (for example, H) or else a *capital* letter followed by a small letter (for example, Cl).

Substances which consist of just *one* type of atom are called **elements**. The element has the same name as the atom. Atoms of the *same element* are usually identical and have the *same mass*; for example, the copper atoms in a piece of copper are all the same. However those atoms from different elements have *different masses*; for example, *hydrogen* is a very light atom, whereas *gold* is a very heavy atom.

It is from these atoms, and combinations of these atoms, that the other two types of particles, ions and molecules, can be made. When atoms react they can do so to form substances containing ions or molecules. These particles are held together by strong forces of attraction called *chemical bonds*. Each atom has a **valency** associated with it (usually a number between 1 to 4) which indicates how many bonds each atom can form.

ATOMIC MASS

Atoms are so very small that a special unit has to be used for measuring their masses. This unit is called the *atomic mass unit* and has the symbol u.

The lightest atom is *hydrogen* with an atomic mass of 1u, whereas *nitrogen* atoms have an atomic mass of 14u. Figure A.22 b) shows an imaginary balance on which one nitrogen atom is balanced by 14 hydrogen atoms. Obviously there is no balance sensitive enough to measure the masses of atoms. However atomic mass *can* be measured in a device called a *mass spectrometer*. This allows the atoms to be ionised before being passed through strong magnetic and electric fields. These fields deflect the ions as they pass through; the heavier the ion, the less it is deflected. Detecting by *how much* an ion has been deflected will give a *measure* of the atom's mass.

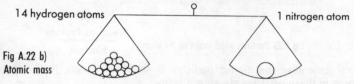

Fig A.22 b)
Atomic mass

The mass spectrometer can give very precise measurements which are quoted in tables, In general, however, we tend to use approximate atomic masses which are often rounded up to the nearest whole number. The following table gives a few examples:

Atom	Symbol	Atomic Mass (u)
Hydrogen	H	1
Magnesium	Mg	24
Oxygen	O	16
Copper	Cu	63.5
Silver	Ag	108

Note: 6×10^{23}u = 1 g. ◄ Atomic number, atomic structure ►

ATOMIC NUMBER

Atoms differ from each other in the number of **protons, neutrons** and **electrons** from which they are made. The *atomic number* is the *number of protons in the nucleus* (which also equals the number of electrons in the atom). The atomic number can be added to the chemical symbol:

$$\text{mass number} \rightarrow A$$
$$\text{atomic number} \rightarrow Z \quad X \leftarrow \text{chemical symbol}$$

The atomic number identifies the type of atom (the name and the way in which it behaves chemically). For example, the carbon atom has an atomic number of 6 (6 protons); all carbon atoms have an atomic number of 6, and they all behave in the same way in chemical reactions. Using the system described above, the carbon atom can be written as $^{12}_{6}C$.

Note: Carbon atoms can have different masses, but will *always* have 6 protons and 6 electrons. ◀ Isotopes ▶

ATOMIC STRUCTURE

The sub-atomic particles that atoms are made of are:

protons ⎫
neutrons ⎭ found in the nucleus
electrons – found orbiting the nucleus

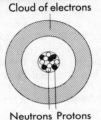

Fig A.23 Particles which make up the atom

The arrangement of the atomic particles is shown in Figure A.23. The differences in these particles are shown below:

sub atomic particle	mass (u)	charge
proton	1	1+
neutron	1	0
electron	very small (1/2000)	1−

▶ PATTERNS FOR ATOMS

If we look closely at the atomic structure of materials, we see that there are several patterns:

- The number of protons in an atom is called the **atomic number**. Atoms have atomic numbers of 1 to 107
- There are *always* the same number of electrons and protons, so every atom is electrically neutral. Charges on the electrons and protons cancel out.
- **Electrons** are arranged in a series of 'shells' around the nucleus. Each shell can only contain a limited number of electrons. The numbers for the first 3 shells are shown below, but after this the arrangement becomes more complex.

    1st shell   maximum 2 electrons
    2nd shell   maximum 8 electrons
    3rd shell   maximum 8 electrons

For example, in the sodium atom, which has 11 electrons, the arrangement is 2 in the 1st shell; 8 in the 2nd shell and 1 in the 3rd shell. Figure A.24 shows the electron configuration (arrangement of electrons in their shells) for sodium as Na:2,8,1.

ATOMIC STRUCTURE

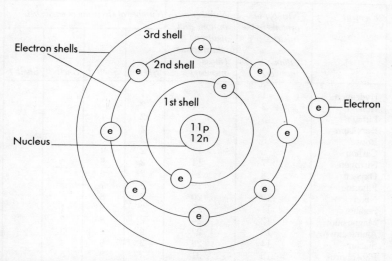

Fig A.24 The electron arrangement of the sodium atom

- The total number of protons and neutrons in the nucleus is called the **mass number** (each proton and neutron has a mass of 1u). There are usually about the same number of protons and neutrons in a nucleus. You can work out the structure of an atom from two numbers:

atomic number = number of *protons* (= number of *electrons*)
mass number = number of *protons* + number of *neutrons*

For sodium, the Atomic number is 11. That is, there are 11 protons and 11 electrons in the sodium atom. The Mass number is 23, which is equal to the number of protons + number of neutrons. Therefore the number of neutrons is equal to the Mass No. − Atomic No., i.e. (23 − 11 = 12). Therefore the sodium atom has 11 protons, 12 neutrons and 11 electrons. The sodium atom can be written as $^{23}_{11}$Na.

The **periodic table** provides us with a complete list of all the atoms that exist in order of **atomic number** (i.e. increasing mass). Each period corresponds to filling an electron shell. Table A.6 shows some common elements. Sometimes atoms of the same type (same atomic number) have different masses; these are called **isotopes**. ◄ Isotopes, Periodic Table ►

ATOMIC STRUCTURE

Table A.6 Atomic structure

Element	Number of protons in the nucleus (atomic number)	Number of protons and neutrons in the nucleus (mass number)	Number of electrons in each shell			
			Shell 1	Shell 2	Shell 3	Shell 4
Hydrogen	1	1	1			
Helium	2	4	2			
Lithium	3	7	2	1		
Beryllium	4	9	2	2		
Boron	5	11	2	3		
Carbon	6	12	2	4		
Nitrogen	7	14	2	5		
Oxygen	8	16	2	6		
Fluorine	9	19	2	7		
Neon	10	20	2	8		
Sodium	11	23	2	8	1	
Magnesium	12	24	2	8	2	
Aluminium	13	27	2	8	3	
Silicon	14	28	2	8	4	
Phosphorus	15	31	2	8	5	
Sulphur	16	32	2	8	6	
Chlorine	17	35.5	2	8	7	
Argon	18	40	2	8	8	
Potassium	19	39	2	8	8	1
Calcium	20	40	2	8	8	2

BACKGROUND RADIATION

We are constantly exposed to **radioactivity**, largely from natural sources. This is referred to as *background* radiation. It arises from small amounts of radioactive atoms present in the air, soil, rocks (particularly granite in this country) and building materials. Background radiation also occurs from the food we eat, mainly due to a radioactive **isotope** of *potassium*. Cosmic rays penetrating the atmosphere also give rise to background radiation. In addition, we receive small doses of radiation arising from *artificial* sources; such as medical treatment (chest and dental X rays), nuclear weapons testing (the dose attributed to this is very small and is dropping as a result of a test ban treaty in the 1960s) and nuclear power (estimated as being very small).

BACTERIA

Bacteria are very small single-celled organisms which reproduce very rapidly (about every 20 minutes) by dividing into two. Bacteria respire either **aerobically** or **anaerobically**, and most are killed at temperatures above 50°C. Bacteria live in the soil, in water and in any dead plants and animals. They are important because they decompose dead plants and animals and release important nutrients for use by plants in the **nitrogen cycle**. Bacteria are also important in breaking down sewage into harmless substances which can be released into rivers. Many bacteria live in your gut where they help to make **vitamin B**. In cows, and other ruminants, bacteria produce **enzymes** to digest the cellulose in plants.

Some bacteria are harmful and cause disease by releasing poisonous substances called *toxins* into the body. These toxins cause diseases such as tetanus and diphtheria. However, a *weakened form* of the bacteria injected into the body will cause the production of **antibodies** which will destroy any more bacteria.

One important effect of bacteria is that they make food go bad and a great deal of money has to be spent on preserving food, such as heating food to kill bacteria.

BALANCED DIET

A balanced diet contains carbohydrates (sugars and starches), protein, fats, minerals, vitamins, fibre (roughage) and water. Figure B.1 shows the main sources and uses of each of these components in the diet.

Type of food	Reason	Source
Carbohydrate	glucose, sucrose, starch – for energy	jams, sweets, bread, potato
Protein	amino acids – for growth and repair of cells	meat, cheese
Fats	fatty acids – storage and energy	butter, oils
Vitamins	A, B, C, D – good health	fresh vegetables and fruit
Minerals	eg iron, calcium – good health	fruit, green vegetables
Roughage	to help bowel movement	vegetables
Water	for all the reactions in the body	fruit and vegetables

Fig B.1

Your daily dietary requirements will vary according to age, pregnancy, illness, and how active a person you are. For example, if you do a lot of exercise you will use up a lot of energy and will need more *carbohydrate* and *fats* which can be broken down to supply energy to your muscle cells. A young person who is still growing will need to eat more foods rich in protein to supply amino acids for the growth of extra body cells to make more tissues and muscles. A labourer on a building site may need more carbohydrate than someone who works in an office. A person who has lost a lot of blood will need to increase their iron intake to form **haemoglobin**, an essential part of the red blood cells.

Children in less well-developed countries often lack one or more essential components of a balanced diet. Children lacking in *protein* suffer from *kwashiorkor*, and are unable to develop proper muscles. Children suffering from a lack of **vitamin A** may suffer from blindness.

BASE

Almost all compounds which contain **ions** can be classified as either **acid**, *base* or **salt**.

Bases are substances which can *neutralize acids*, they react with acids to produce salts. Bases include metal oxides, e.g. magnesium oxide; metal hydroxides, e.g. calcium hydroxide and metal carbonates, e.g. zinc carbonate. Most bases are insoluble in water; those which *do* dissolve are called **alkalis**.

BATTERY

◀ Cell − electrical ▶

BAUXITE

Bauxite is the main ore of aluminium (the most common metal in the Earth's crust). Bauxite is a rocky material often reddish in colour which has a very high proportion of aluminium oxide (Al_2O_3).

Aluminium is obtained from bauxite in two stages. In the first stage the bauxite is crushed and treated with *sodium hydroxide* to separate the aluminium oxide (a white powder) from its impurities. In the second stage the metal is extracted from aluminium oxide by **electrolysis**.

BETA RADIATION

When the nucleus in radioactive materials breaks down it can emit three different types of radioactivity: **alpha**, *beta* and **gamma**.

Beta (β) particles are very fast-moving **electrons** and travel further through the air than alpha particles and are more difficult to stop. They are able to penetrate skin, but can be stopped by thin sheets of metal and are deflected by a magnetic field (Fig. B.2).

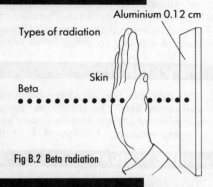

Fig B.2 Beta radiation

BIMETALLIC STRIP

◀ Thermostats, feedback systems ▶

BIODEGRADABLE

Biodegradable means capable of being broken down by bacteria. However, substances such as **plastics**, which are widely used in packaging, are not biodegradable. Research is being carried out to make biodegradable plastics by including short chains of sugar into the plastic molecule. Bacteria will then break down the sugar and leave very short chains which will degrade in the environment.

BIOLOGICAL CONTROL

Biological control is when animals are introduced into the food chain to control another animal or plant which has become a pest. Sometimes a *natural predator* of the pest is introduced to feed on the pest; for example, the red spider mite is a very common insect pest which damages plants in greenhouses. Instead of spraying with insecticide, a gardener can introduce another insect which is a predator of the red spider and which eats about 20 red spiders a day!

One of the *disadvantages* with this method of controlling pests is that it is a much slower process than using chemical pesticides. Another disadvantage is that the predator can itself become a pest, and may select a different source of food to the intended one. Some of the *advantages* of biological control are:

1 there is no pollution
2 there is no danger to the rest of the food chain in terms of accumulation of pesticide in the **carnivores**

3 dangerous chemicals are not used and therefore insects which are resistant to the pesticides do not develop.

BIOMASS

The term biomass refers to the amount (mass) of living material in a **food web** or **ecosystem**. It can be determined by weighing all the living things at each **trophic level; producers, herbivores, carnivores**, etc. However, the biomass should be weighed over a period of time, for example, a year, as the amount of material will vary according to the season. ◄ Pyramid of biomass ▶

BIOSPHERE

This describes the part of the Earth and the **atmosphere** in which all living organisms live. On Earth the biosphere consists of land and water environments. Organisms are usually adapted to a *particular* environment and live in a specialised **habitat**. ◄ Ecosystem ▶

BIOTIC FACTORS

Biotic factors are the living community of animals and plants in an **ecosystem**. In other words, the green plants or **producers** which make food by **photosynthesis**, the **consumers** which obtain food from the plants or other animals, and the **decomposers** which break down and decompose the dead animals and plants.

BISTABLE CIRCUITS

The main feature of a bistable circuit is that whatever happens to the input, the output has only one of two stable states, each of which is achieved by **feedback**. In Figure B.3, if input A is logic high and input B not connected, then indicator A is off and the B input is at logic 0. Similarly, if input A is not connected and B is at logic 1, the output of B is at logic 0 and to A is at logic 1.

In these circuits a connection between output and input *reinforces* information. If the output of B is fed back to the input of A, a bistable system is created. Input A goes high, input B is not yet connected; output A goes low, output B goes high and reinforces (feedback) the high input to A. If the original connection to A is removed, the message is retained: in other words the system has a *memory*.

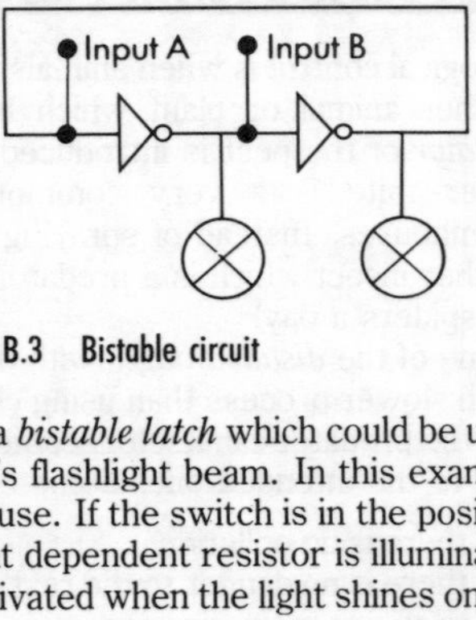

Fig B.3 Bistable circuit

Two NOR-gates can be used to produce a *bistable latch* which could be used as a burglar alarm activated by the burglar's flashlight beam. In this example (Fig B.4), only one of the two outputs is in use. If the switch is in the position shown, there is no output, whether the light dependent resistor is illuminated or not. When the switch is up, the bell is activated when the light shines on the

LDR and remains sounding until the switch is moved again. This is a 'latch' system, keeping one steady desired state. ◄ Feedback ►

Fig. B.4 A bistable latch burglar alarm

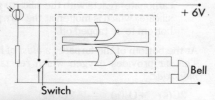

BLAST FURNACE

There are two basic techniques for extracting a metal from its ore; either **reduction**, using heat energy and carbon as the reducing agent, or **electrolysis**, using electrical energy. In either case the problem is the same, namely to reduce the metal **ion** to a metal **atom**. The technique that is chosen depends on cost and the reactivity of the metal. Reduction is carried out in a *blast furnace* (Fig B.5).

▶ IRON EXTRACTION

One metal to be extracted from its ore by reduction is *iron*. Iron ore, or haematite, contains iron oxide, and is 'smelted' in a blast furnace. Once the furnace is started it operates as a *continuous process*; the raw materials being added at the top and the molten iron and molten waste materials being run off at the bottom (Fig B.6).

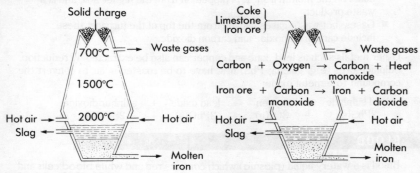

Fig. B.5 The blast furnace

Fig B.6 Section through the blast furnace used to extract iron from iron ore

The raw materials

These are added at the top of the furnace. They consist of:
- *iron oxide ore*, which will be reduced to iron
- *coke*, which provides the reducing agent
- limestone, which is added to remove the waste rock in the iron ore. Iron ore contains a lot of other impurities (mainly silica), which would soon clog the furnace and have to be removed, needing the furnace to be shut down and allowed to cool – a costly procedure. The limestone (calcium

carbonate) reacts with the silica to produce a glassy material (calcium silicate) which is molten at the furnace temperature and runs to the bottom to be tapped off.

At the bottom of the furnace, a *blast* of hot air is forced into the molten mass. This air provides oxygen which reacts with the carbon to produce *carbon monoxide*:

$$2C(s) + O_2(g) \rightarrow 2CO(g)$$

The reaction

Carbon monoxide is a powerful reducing agent and reduces the iron oxide to iron, which is molten at the temperature of the furnace:

carbon monoxide + iron oxide → iron + carbon dioxide
$$3CO(g) + Fe_2O_3(s) \rightarrow 2Fe(l) + 3CO_2(g)$$

At the same time the high temperature of the furnace causes the calcium carbonate to be converted into calcium oxide, which then reacts with the silica to produce *molten calcium silicate* (slag).

The products

- Iron: molten iron is very dense so travels down through the furnace and is tapped off at the bottom hole into large moulds called 'pigs'. The iron that is produced is called *pig iron*.
- Slag: this is the molten calcium silicate: it is less dense than the iron so it floats on the molten iron. It is tapped off from the higher hole and is a waste product.
- Gases: hot waste gases escape from the top of the furnace; these include carbon monoxide and carbon dioxide.

Other metals such as lead, zinc and copper can also be extracted by reduction. Sulphide ores (e.g. galena; PbS) first have to be roasted in air to convert the compound to a metal oxide:

lead sulphide + oxygen → lead oxide + sulphur dioxide
$$2PbS(s) + 3O_2(g) \rightarrow 2PbO(s) + 2SO_2(g)$$

BLOOD

Blood is a watery liquid (**plasma**) which contains red and white blood cells and

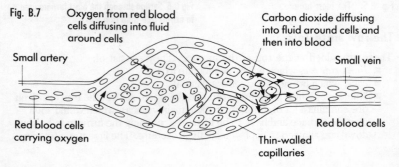

Fig. B.7

tiny particles called **platelets**. Adults have about five to six litres of blood in their body. There are two main functions of blood. One is to maintain a *constant internal environment* for all the cells in the body by providing oxygen and food and removing waste products. The second main function is to *transport* many different substances around the body in the **circulatory system**. ◀ Breathing ▶

BLOOD SYSTEM

The blood system is a continuous series of tubes inside your body which carry blood to all the different parts of you. The main tubes are either **arteries**, carrying blood away from the heart, or **veins** carrying blood towards the heart. Blood constantly flows around in one direction pumped by the **heart**, a powerful muscle. The blood vessels divide up into smaller and smaller tubes and eventually form tiny thin-walled **capillaries**.

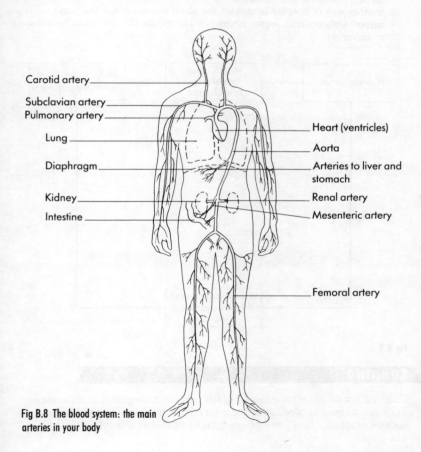

Fig B.8 The blood system: the main arteries in your body

BODY TEMPERATURE

In adult humans body temperature is about 37°C, but this can vary during the day for an individual person. Young children have slightly higher body temperatures than adults. You are able to control your body temperature to keep it constant by various methods (Fig B.9).

If you are too hot you *lose heat* by:

1. the evaporation of sweat, which has a cooling effect on the body, and
2. by losing heat from the blood in tiny vessels just under your skin, which widen (**vasodilation**) to allow heat loss by radiation.

If you are too cold you can *retain heat* by:

1. increasing the amount of clothing which traps air, a poor conductor of heat, around your skin. In other mammals and birds this function is carried out by fur and feathers.
2. another way of keeping heat is for the blood vessels under the skin to narrow and constrict (**vasoconstriction**) to reduce the amount of heat lost by radiation.

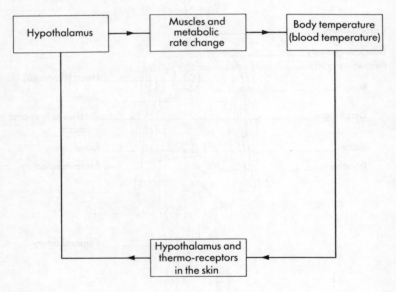

Fig B.9

BONDING

Chemical bonds are strong forces which hold atoms together in substances. They are formed as a result of the interactions of electrons which orbit the nucleus of atoms. There are different types of chemical bonds:

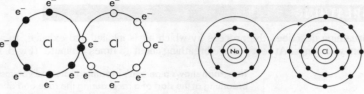

Each chlorine atom now appears to have a full outer shell: shared electrons

Fig B.10 a)   Fig B.10 b)

1. **Covalent bonds** – these are formed when atoms share their electrons and are found in molecules (mostly between non-metal atoms) (Fig B.10a)).
2. **Ionic bonds** – these are formed when one atom loses an electron to another atom resulting in **ions** of opposite charges being formed. This happens mostly between a metal atom (which gives an electron) and a non-metal atom (which accepts an electron) (Fig B.10b)).
3. *Metallic bonds* – in metals the atoms 'float' in a sea of electrons.

Chemical bonds are strong and have to be broken before one substance will react with another. This is the reason we often have to heat substances to make them react.

BRAIN

The brain is a very large part of the nervous system. Its function is to coordinate many of the body's activities. It receives input from the sensory organs and sends motor impulses to the muscles and glands in the body. The brain is an important store of information so that animals can learn from past experiences.

The main parts of the brain and their functions are shown in Figure B.11.

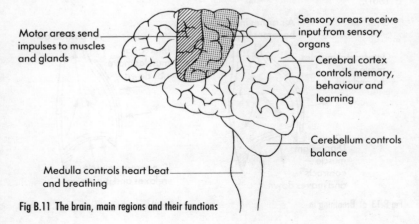

Fig B.11 The brain, main regions and their functions

BREATHING

Breathing describes the process by which air is inhaled and exhaled, to and from your lungs. At rest you are breathing about 15 times a minute. (Fig B.12)

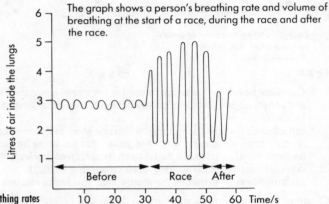

Fig B.12 Breathing rates and lung volume

Try putting your hands over your ribs and taking a deep breath. You should be able to feel your chest cavity getting larger as you breathe in. The *intercostal muscles* between your ribs contract to pull your ribs up and out. At the same time the *diaphragm* muscle at the base of your chest flattens so that your chest cavity is made larger. Air outside your chest cavity is at greater pressure than air inside your chest. This difference in pressure causes air to rush into your lungs (Fig B.13a). When you breathe out, the intercostal muscles and diaphragm contract, making your chest cavity smaller, therefore increasing the pressure in the lungs. As a result, the air is pushed out (Fig B.13b)).

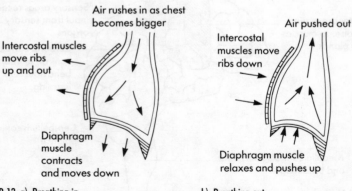

Fig B.13 a) Breathing in   b) Breathing out

BREATHING

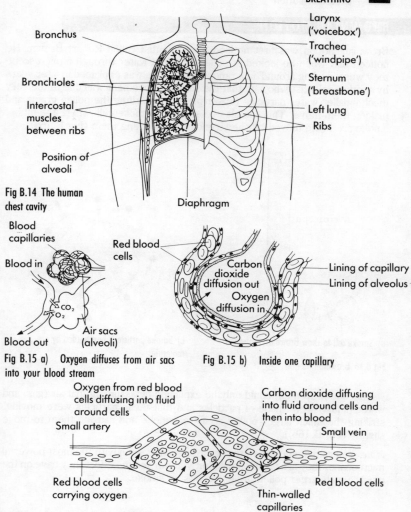

Fig B.14 The human chest cavity

Fig B.15 a) Oxygen diffuses from air sacs into your blood stream

Fig B.15 b) Inside one capillary

Fig B.15 c) Oxygen and carbon dioxide transport

Your *lungs* are basically two sponge-like structures in your chest which fill up with air (Fig B.14). Oxygen diffuses over the moist surface of the air sacs (or *alveoli*) which in turn help oxygen diffuse into the blood in the capillaries. Here it combines with **haemoglobin** in the red blood cells, to make a new substance called *oxyhaemoglobin* (Fig B.15a)). The blood is pumped by the **heart** muscle to the rest of the body through **arteries** and eventually **capillaries** (Fig B.15b)). The oxygen diffuses into your cells (Fig B.15c)), and carbon dioxide from the cells diffuses into your blood and is carried back to the lungs.

BROWNIAN MOTION

Brownian motion was first noticed by a scientist called Robert Brown. He noticed that when he looked at pollen grains in water through a microscope they were 'jiggling around' in a random way. This was explained in later years by another scientist who said that the strange movement was due to the very much smaller water particles (water molecules) *hitting* the pollen grains and making them move. This can also be seen in a 'smoke cell' where the bits of smoke are being moved by the air molecules striking them (Fig B.16a)).

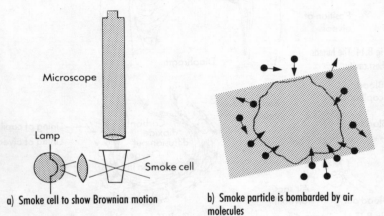

a) Smoke cell to show Brownian motion

b) Smoke particle is bombarded by air molecules

Fig B.16 Brownian motion

This odd movement could only be explained by assuming that air (gas) and water (liquid) were made of particles and that these particles were moving, striking the much larger pollen grains or smoke and causing them to move slightly (Fig B.16b)).

Note: You cannot see the water or air molecules even with the most powerful microscope, they are far too small, but you *can* see the effect they have on the much, much larger pollen grains or 'bits' of smoke.

CANCER

One of the commonest causes of death in Western society is from some form of *cancer*. A cancer is caused by a group of **cells** in the body which begin to grow abnormally and destroy healthy cells. It is thought that environmental factors such as smoking, alcohol, and radiation can increase the risk of cancer developing. Cancers can be caused in most parts of the body, for example the lungs, the large intestine, the skin, and the breasts. Prevention of cancer may be helped by reducing the amount of fats in the diet and by increasing the intake of fresh vegetables.

CARBOHYDRATES

These are a family of compounds which include **sugars** and **starches**. All carbohydrates contain carbon, hydrogen and oxygen. The name carbohydrate is obtained from the terms *carbo-* meaning carbon, and *hydrate*, meaning hydrogen and oxygen present, in the proportions they are in water.

Examples of carbohydrates are glucose (a sugar), formula $C_6H_{12}O_6$ and sucrose (a sugar), formula $C_{12}H_{22}O_{11}$.

Carbohydrates are manufactured by plants, during the process of **photosynthesis**, to produce simple sugars. The plant then converts these into more complex molecules, such as starch and cellulose (both **polymers**).

When we eat plant material as food, it provides us with a source of carbohydrate which in turn is broken down inside our bodies to provide energy. Foods high in carbohydrate content are bread, rice and vegetables such as potato.

CARBON

Carbon is an **element** (it contains only one type of atom), chemical symbol C, that has two crystalline forms.

a) diamond – one of the hardest substances known;
b) graphite – in this form carbon will conduct electricity.

CARBON

Carbon is also found in coal, coke and charcoal. Carbon is a good **reducing agent**, in that it will 'grab' oxygen from many other compounds. For example, it will remove oxygen from iron oxide, leaving iron. ◀ Allotropes, Blast furnace ▶

CARBON COMPOUNDS

Carbon is an unusual element, since it can form a very large number of different substances which consist of long carbon chains, or rings of carbon atoms. It is this ability for carbon atoms to *link with each other* that make it almost unique (*silicon* has a similar ability). There is a branch of chemistry called organic chemistry, devoted to the study of just these *compounds*:

- All living things consist of carbon compounds; proteins, starches and cellulose.
- Sugars consist of carbon chains or rings;
- All the substances found in crude oil consist of rings or chains of carbon compounds;
- All man-made polymers (plastics) consist of long carbon chains.

CARBON CYCLE

Carbon is breathed out, as **carbon dioxide**, by all animals and plants. Whenever **fossil fuels** (such as coal, oil or gas) are burnt, carbon dioxide is also *released* into the **atmosphere**.

Green plants *take in* carbon dioxide during the day time, and combine the carbon dioxide with water to make **carbohydrates**. This process is known as **photosynthesis** and it releases oxygen as a waste product.

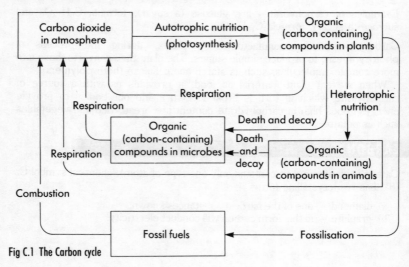

Fig C.1 The Carbon cycle

The plants are eaten by animals and the carbon, in the form of the carbohydrates, proteins and fats, is used to make the cells of the animals. As the animals **respire**, the carbohydrates are broken down to form carbon dioxide and water. The carbon is released as carbon dioxide to the atmosphere, completing the carbon cycle (Fig C.1).

CARBON DATING

All living things contain a large amount of carbon in them. Most of the carbon atoms are of the **isotope** carbon-12, but a small proportion are of the radioactive isotope, carbon-14. This proportion of carbon-12 to carbon-14 is the same for all living things whilst they are alive. When the organism dies, the amount of carbon-14 decreases (half-life 5736 years). By measuring the amount of radioactive carbon left, one can *date* the item by reference to the **half-life** curve. This technique of carbon-14 dating was recently used to date the Turin Shroud.

CARBONATES

Carbonates belong to a group of substances called **salts**. All carbonates contain the **carbonate ion** which is negative and a positive **ion** (metal).

CARBONATE ION

Examples of the carbonate ion, CO_3^{2-}, are zinc carbonate ($ZnCO_3$), copper carbonate, ($CuCO_3$) and calcium carbonate, ($CaCO_3$).

Most carbonates are not soluble; the exceptions being sodium, potassium and lithium. *Metal carbonates* can be thought of as **bases**; they will react with acids to produce salt and water and carbon dioxide.

General pattern of the reaction

acid + carbonate → salt + water + carbon dioxide

Examples of this reaction are:

sodium carbonate + nitric acid → sodium nitrate + water + carbon dioxide

$Na_2CO_3 + 2HNO_3 \rightarrow 2NaNO_3 + H_2O + CO_2$

calcium carbonate + hydrochloric acid → calcium chloride + water + carbon dioxide

$CaCO_3 + 2HCl \rightarrow CaCl_2 + H_2O + CO_2$

This is a general pattern of reaction for all carbonates and, as many rocks are carbonates, it can be used as a *test* to help identify rocks. For example, limestone, marble and chalk are all mainly calcium carbonate and will therefore react with hydrochloric acid. When a small amount of acid is placed on the surface, the rocks 'fizz' and give off carbon dioxide.

Note: there is an exception; sulphuric acid will *not* react well with calcium carbonate rock because, during the reaction, a layer of calcium sulphate builds up on the surface preventing any further reaction.

CARBON DIOXIDE

Carbon dioxide's properties are as follows:
- colourless
- odourless
- more dense than air
- dissolves slightly in water to produce a weakly acidic solution.

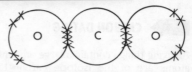

Fig C.2 Structure of carbon dioxide

Carbon dioxide is a gas made up of **molecules** which contain one atom of carbon and two atoms of oxygen (formula CO_2) (Fig C.2).

Carbon dioxide is the only gas that will turn limewater cloudy or 'milky'; this can be used as a test for the gas.

Carbon dioxide is given off when **fossil fuels** and plant materials are burnt and when carbonates are heated. It is also released into the atmosphere by the respiration of animals and made use of by plants during **photosynthesis**.

CARNIVORES

Carnivores are the **consumers** in a **food chain** which obtain their energy by feeding on the **herbivores**. Some carnivores obtain their energy from other carnivores, and these are described as *tertiary* (or third level) consumers; for example, a hawk or fox is a predator which feeds on other carnivores in the food chain.

CATALYSTS

A catalyst is a substance that changes the rate of a chemical reaction; the catalyst itself remains unchanged at the end of a reaction and can be re-used. Catalysts are often used to speed up reactions, but they can also be used to slow them down.

GAS REACTIONS AND CATALYSTS

One way in which catalysts are thought to work in reactions involving gases is by the surface of the catalyst providing *sites* where the reacting molecules can 'meet'. The **transition metals** are often used as catalysts in this way in industrial processes.

Manufacture of ammonia

$$N_2 + 3H_2 \rightleftharpoons 2NH_3$$

Iron is used as a catalyst; nitrogen and hydrogen are *adsorbed* onto the surface where they come into contact and react. In the gas state they are often moving too fast, so that when they collide they bounce off each other *without* reacting.

Manufacture of sulphuric acid

One stage involves the production of sulphur trioxide from sulphur dioxide:

$$2SO_2 + O_2 \rightleftharpoons 2SO_3$$

Vanadium (V) oxide is the catalyst and works in a similar way to that described above.

Pollution control on cars

Petrol engines in cars burn petrol (a hydrocarbon fuel) and produce carbon dioxide and water as waste products. In addition, carbon monoxide and some oxides of nitrogen are produced, which pollute the atmosphere. The car exhaust systems in California (where the problem of pollution from cars is a very serious one) are fitted with a metal catalyst with a large surface area. As the hot exhaust gases pass over the catalyst, the pollutant gases are converted into carbon dioxide and nitrogen.

Antioxidants

Certain chemicals can be added to foods, such as crisps, to slow down the natural oxidation of foods, which would result in loss of flavour and decay. These chemicals (*antioxidants*) are acting as catalysts. Here they are *reducing* the rate of a chemical reaction.

 ENZYMES

These are biological catalysts and enable chemical reactions to take place in living things. For example, the enzyme salivary **amylase** aids the breakdown of starch to sugar.

CATHODE

The cathode is the name given to a negative **electrode** either in an electrolysis cell or an electrical cell (battery).

 IN ELECTROLYSIS

The cathode and **anode** (the positive electrode) are attached to an electrical source (e.g. battery or power-pack) and dipped into an electrolyte (liquid or solution containing ions). A current flows through the electrolyte when positive ions (**cations**) are attracted to the cathode. At the cathode these ions are converted to atoms:

$$M^+ + e^- \rightarrow M$$

The cathode must conduct electricity and is usually carbon (graphite) or a metal.

IN AN ELECTRICAL CELL

When two different metals (which act as electrodes) are placed in an electrolyte and are connected together externally, for example with a wire, they produce an electric currrent (a flow of electrons). The cathode (the *negative* electrode) will be the metal which has the greatest tendency to lose electrons (higher in the reactivity series). In the dry cells (batteries) that we buy, the cathode is often made of zinc.

◄ Electrolysis, Cell — electrical ►

CATHODE RAY OSCILLOSCOPE (CRO)

A CRO can be used as a visual voltmeter for measuring **voltage**. A bright spot is produced on the oscilloscope screen by a beam of electrons. The position of the spot can be altered by the voltage across the CRO. When the time-base control is adjusted, the dot moves across the screen and draws a visual graph of the voltage against time.

Figure C.3 shows the waveform of a DC supply, and an AC supply.

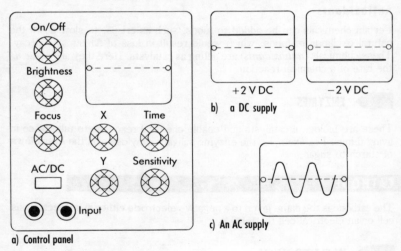

a) Control panel

b) a DC supply

c) An AC supply

Fig C.3 Cathode ray oscilloscope

CATION

Substances which contain **ions** (charged particles) and which are electrically *neutral* (they contain equal numbers of positive and negative charges).

Examples of ionic compounds are M^+X^-, or $N^{2+}Y^{2-}$. Those ions which have a positive charge are called cations, whereas those which have a negative charge are called **anions**.

Metal atoms tend to form cations, whereas non-metal ions tend to form anions.

- Examples of **some common cations** are shown below:

Name of ion	symbol	charge on ion
copper	Cu^{2+}	2+
sodium	Na^+	1+
calcium	Ca^{2+}	2+
aluminium	Al^{3+}	3+
ammonium	NH_4^+	1+

CELL – BIOLOGICAL

Cells are the basic units which make up all living organisms (plants and animals). A cell is basically a very tiny fragment of *cytoplasm* surrounded by a cell membrane (Fig C.4). The cell's activities are controlled by a nucleus which contains genetic material in the form of **chromosomes**.

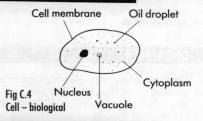

Fig C.4 Cell – biological

CELL – ELECTRICAL

The difference in tendency of metals to form **ions** can be very useful. If a pair of different metals is placed in a solution containing ions and the metals are linked by a wire, then **electrons** will flow through the wire. This means that a current is flowing through the wire, and there is a **voltage** between the two metals. This arrangement is called a *simple cell*, as shown in Figure C.5.

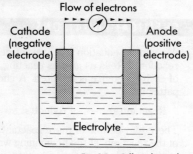

Fig C.5 Cell – electrical

All cells contain three things:
- a + terminal (the positive electrode or **anode**)
- a − terminal (the negative electrode or **cathode**)
- a solution containing ions through which electricity can pass (**electrolyte**).

The voltage that is produced between the two metals generally depends on their relative positions in the **reactivity series**. If the metals are *far apart* in the reactivity series, then a *large* voltage is produced; if the two metals in the pair are *close together*, then a *small* voltage is produced. For example:

Metal pair	Voltage produced
magnesium/copper	large voltage
iron/zinc	small voltage

Dry cells (batteries) that you can buy consist of metal pairs in an electrolyte, except that one of the metals is replaced by carbon (in the form of graphite). In normal dry cells, the electrolyte is a weak acid, in the form of a paste. Alkali batteries have electrolytes which are alkalis (e.g. potassium hydroxide), the positive electrode often being a mixture of carbon and manganese oxide.

CELSIUS SCALE (°C)

This is a scale of temperature which has 100 degrees (divisions) between the lower fixed point of 0°C and the upper fixed point of 100°C. The *lower fixed point* is the temperature at which ice melts and changes to water. The *upper fixed point* is the temperature at which pure water boils and changes to steam.
◀ Kelvin ▶

CERAMICS

Ceramics are materials which are made from clays and have been heated at some stage of their manufacture, such as pottery, bricks, etc. They consist of giant structures of silicon and oxygen atoms (*silicates*).

Ceramics are useful because they are: strong (in compression); hard; non-conductors of heat and electricity; and are chemically inert. They are also relatively easily cleaned.

CHEMICAL EQUATIONS

A chemical equation is a way of describing what is happening in a chemical reaction. It tells us *what* is reacting; *what* products are formed and *how much* of each substance is required. A chemical reaction always follows the general pattern:

Reactant(s) → Product(s)

There may be one or more reactants and one or more products, but the mass of the reactants at the beginnning will be the same as the mass of the products at the end. In an equation representing the reaction, you will have the same number of atoms (represented by their symbols) on the left as you will have on the right hand side. All that is happening in a chemical reaction is a rearrangement of these atoms.

 WRITING CHEMICAL EQUATIONS

It may help to demonstrate with actual examples:

Example 1

Sodium reacts with chlorine to form sodium chloride:

- Step 1 Write the equation in words:

    sodium + chlorine → sodium chloride

CHEMICAL EQUATIONS

- Step 2 Write each substance as a formula:

    Na + Cl$_2$ → NaCl

    Sodium is an element (Na); reactive gaseous elements are diatomic (Cl$_2$); sodium chloride is Na$^+$Cl$^-$.

- Step 3 Imagine the reaction as particles:

    Na + Cl$_2$ → NaCl

    (Na) + (Cl)(Cl) → (Na)(Cl)

- Step 4 Balance the equation:

    2 Cl on the left, so must have 2Cl on the right.
    The only way is to obtain this is to have 2NaCl on the right, as follows:

    2NaCl   (Na)(Cl)
               (Na)(Cl)

    Na + Cl$_2$ → 2NaCl

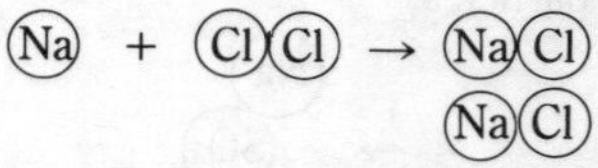

    (Na) + (Cl)(Cl) → (Na)(Cl)
                             (Na)(Cl)

    Now we must have 2 Na on the left to balance the equation:

    2Na + Cl$_2$ → 2NaCl

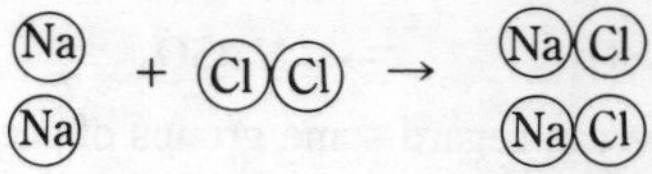

    (Na)   + (Cl)(Cl) → (Na)(Cl)
    (Na)   (Na)(Cl)

    4 atoms   4 atoms

The equation is balanced. *Remember* : when balancing an equation NEVER change the actual formulae.

Example 2

Magnesium reacts with hydrochloric acid to produce magnesium chloride and hydrogen:

- Step 1 Write the equation in words:

    magnesium + hydrochloric acid → magnesium chloride + hydrogen

- Step 2 Write each substance as a formula:

    Mg + HCl → MgCl$_2$ + H$_2$

- Step 3 Imagine the reaction as particles:

    (Mg) + (H)(Cl) → (Cl)(Mg)(Cl) + (H)(H)

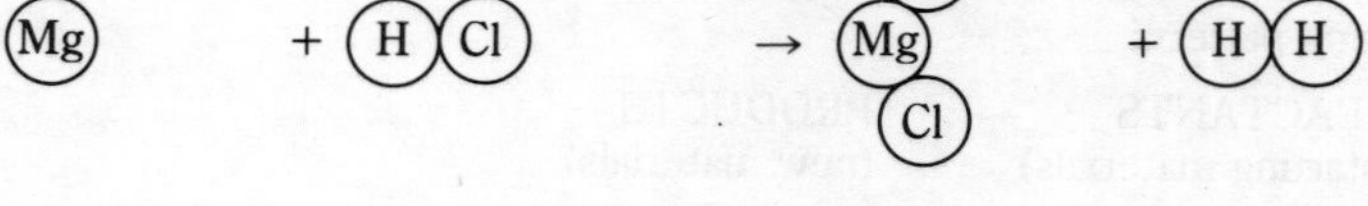

- Step 4 Balance the equation:

  Mg + 2HCl → MgCl$_2$ + H$_2$

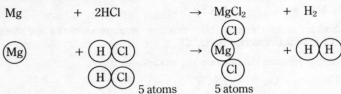

Example 3

Sulphuric acid neutralises sodium hydroxide:

- Step 1 Write the equation in words:

  sodium hydroxide + sulphuric acid → sodium sulphate + water

- Step 2 Write each substance as a formula:

  NaOH + H$_2$SO$_4$ → Na$_2$SO$_4$ + H$_2$O

- Step 3 Imagine the reaction as particles:

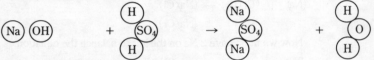

- Step 4 Balance the equation:

  2NaOH + H$_2$SO$_4$ → Na$_2$SO$_4$ + 2H$_2$O

This example also shows that we can regard some groups of particles as one unit which is not usually changed in a chemical reaction, e.g. the sulphate ion SO$_4^{2-}$. Other examples where we can do this are OH$^-$ (hydroxide ion); NO$_3^-$ (nitrate ion); and sometimes CO$_3^{2-}$ (carbonate ion in displacement reactions) and HCO$_3^-$ (hydrogen carbonate ion). More unusual ions are SO$_3^{2-}$ (sulphite ion) and NO$_2^-$ (nitrite ion).

CHEMICAL FORMULA

A chemical formula for a compound shows the ratio of atoms present in that compound whether they are joined by Ionic or Covalent bonds. Each atom has a combining power which is called the Valency. The valency of an atom depends on the number of Electrons in its outer shell and hence its position in the Periodic table. ◀ Valency ▶

CHEMICAL REACTIONS

Chemical reactions are interactions between particles (**atoms, molecules** or **ions**) which involve the 'breaking' and 'making' of chemical **bonds**. There is a general pattern:

REACTANTS → PRODUCTS
(starting materials)   (new materials)

New substances are always formed. These are ones which have a different set of properties to those of the reactants.

Mass stays the same. In any chemical reaction the total mass of the *products* is always the same as the total mass of the *reactants*. The reason for this is that chemical reactions only involve the *rearrangements* of the particles involved.

$$A\ B\ +\ C\ \rightarrow\ A\ C\ +\ B$$

You can think of this in the same way as dismantling a Lego model of a house and a garage and using the same bricks to build a small factory and storehouse. This allows us to write equations and to make calculations for particle interactions, e.g. the mass of reactants needed to produce 1kg of a certain product (this is invaluable in the chemical industry). *Energy* is always involved in a chemical reaction:

Activation energy: not all substances will react with each other but many do. Some combinations of reactants need a 'push' to get them going – usually in the form of heat – this is referred to as the activation energy. During a reaction sometimes heat is given out (**exothermic**) and sometimes heat is taken in (**endothermic**).

 RATE OF REACTION

Chemical reactions take place at different *rates*, and this rate (speed) can be measured. Often ways are needed to *control* the rate of these reactions, either to speed them up or to slow them down (this is important in the chemical industry). ◀ Catalysts ▶

CHEMICAL SYMBOL

There are approximately 100 different elements (different atoms) and each one has a chemical symbol.

Each chemical symbol is either one capital letter (often the first letter of the name) or a capital letter followed by a small letter. Here are some examples: carbon, C; chlorine, Cl; zinc, Zn; sodium, Na; potassium, K; oxygen, O.
◀ Periodic table ▶

CHLORIDE

Chlorides belong to a group of substances called **salts**. All chlorides contain the *chloride ion* which is negative, and a positive ion (metal or ammonium). Examples of compounds which contain the chloride ion (Cl^-) are: sodium chloride, NaCl; zinc chloride, $ZnCl_2$; magnesium chloride, $MgCl_2$; ammonium chloride, NH_4Cl.

Most chlorides are soluble in water; the exceptions are silver and lead. Small amounts of chlorides (mostly calcium and magnesium) can be found in tap water, whereas sea water contains much larger quantities of chlorides (as sodium and potassium salts).

CHLORINE

Chlorine is an **element** (contains only one type of atom) with chemical symbol Cl.

It is a green, poisonous gas which contains chlorine molecules (Cl_2). It is quite soluble in water, with which it reacts to produce an acidic solution that also has a bleaching action. In fact the smell you get from bleach is that of chlorine gas which is given off by the bleach. Chlorine also acts as an antiseptic (kills germs) and so is added to swimming pools in the form of a solution.

CHLOROFLUOROCARBONS

Chlorofluorocarbons, known as CFCs, are a group of compounds which contain carbon, chlorine and fluorine atoms, and which have boiling points at just below room temperature. They are not flammable and have low toxicity. This makes them ideal for use in aerosols and refrigerators. They are also used in the manufacture of plastic packaging. Unfortunately, they have been linked to damage of the **ozone layer** which protects us from harmful **ultra-violet rays**.

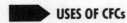

USES OF CFCs

Aerosol use

CFCs are used as the propellant gas in some aerosols; their use has now been banned from 1992.

Refrigeration use

CFCs are used in the refrigerants; they are the liquids which circulate in the black pipes at the back of the fridge. They take the heat away from the inside of the refrigerator.

Plastic packaging manufacture

CFCs are used as 'blowing' or foaming agents to make rigid or flexible plastic foam. Some of the boxes used by fast-food chains to hold hamburgers or for egg packaging use CFC rigid foam.

There are about a dozen different CFCs in use today, some of which are more harmful to the ozone layer than others; companies are beginning to use the less harmful CFCs or to look for alternatives. The European governments have agreed to eliminate the use of CFCs by the year 2000.

CHROMATOGRAPHY

Chromatography is a technique for detecting the parts of mixture by separating them.

PAPER CHROMATOGRAPHY

A drop of solution of the mixture to be separated is placed on a type of blotting paper (called *chromatography paper*) and then dipped into a solvent. As the **solvent** soaks up through the paper it carries the mixture with it, but because different substances dissolve at different rates, the solvent carries some parts of the mixture further than others, so separating them. This can be seen if a dot is made with a black felt-tip pen on chromatography or filter paper and then dipped into a solvent such as water. The colours which make up the black (e.g. purple, red, blue, green, etc.) will separate. This technique can be used to separate a mixture of amino acids or proteins.

GAS CHROMATOGRAPHY

The mixture is injected into a long, coiled tube which contains a solid material, called a diatomaceous earth. The mixture is carried along by a gas which is moving slowly through the tube. As the gas carries the mixture through, it separates, because it carries some substances through *faster* than others. This is how different substances can be detected in the blood (e.g. alcohol) because the operator knows how long it takes for alcohol to travel through the tube.

CHROMOSOMES

Normal human body cells have 23 pairs of chromosomes in the nucleus; every 23 pairs carries instructions for the whole human body. Each chromosome holds the information for many chemical reactions and is made of complex molecules of DNA (**deoxyribonucleic acid**). For example, the instructions for making the enzyme salivary amylase and the Rhesus blood antigens are thought to be part of the same chromosome. ◄ Gene ►

CIRCUIT SYMBOLS

Some of the conventional symbols for electrical circuit diagrams are reproduced in Figure C.6.

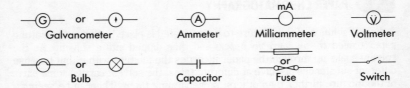

Fig C.6 Electrical circuit symbols

CLIMAX COMMUNITY

This term describes a stable and balanced **community** of animals and plants which occupy an area (Fig C.7). The community is in equilibrium with the environment until some change, such as a change of climate, takes place.

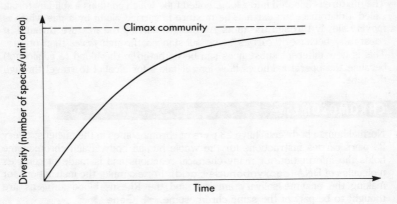

Fig C.7 Climax community

CLOUDS

Clouds are lots of small drops of water formed by **water vapour** condensing when warm air moves up to a higher level where the air is cooler.

HOW CLOUDS ARE FORMED

1. When air passes over water, the air picks up water vapour which has evaporated from the surface.
2. When the air has absorbed the maximum amount of water vapour which it can hold at that temperature, it is said to be *saturated*. Colder air can hold less water vapour than warmer air.
3. The energy from the Sun heats up masses of air and causes the air molecules to spread out, so the air becomes less dense and rises up.

COEFFICIENT OF EXPANSION

4 As the air rises, it cools down, and any water vapour in the air begins to condense and form clouds.
5 When the cloud rises higher, the air in the cloud is saturated with water vapour, which is then released as rain.

The process is shown diagrammatically in Figure C.8.

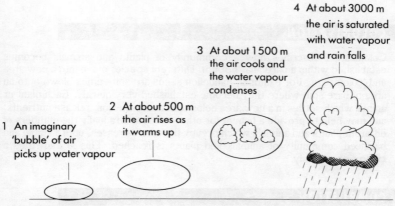

Fig C.8 How clouds are formed

▶ CLOUD TYPES

Name	Height	Description
cirrus	7 to 10 km	wispy white threads formed from ice crystals
cumulus	3 to 5 km	thick white clouds with flat base and rounded tops
nimbus	2 to 3 km	thick grey clouds usually rain or snow produced
stratus	0.5 to 2 km	continuous sheet of low cloud formed in calm weather may thicken and produce rain

COCHLEA

◀ Ear ▶

COEFFICIENT OF EXPANSION

◀ Expansion ▶

COKE

Coke is made by heating coal in the absence of air so that it does not burn. This drives off substances contained in the coal which tend to produce a lot of smoke when they burn. The result is coke, which burns with almost no smoke. Coke is used as a reducing agent in the **blast furnace**.

COLONISATION

Colonisation describes how a **community** of plants and animals becomes established within a particular **habitat**. Different **species** move into a new area and 'colonise' it. For example, dandelion seeds are sometimes blown onto an area of bare soil, where they become established very quickly. Each plant or animal which settles in a new area helps to stabilise the soil, release nutrients, and may help to provide a habitat for other species. Gradually the numbers of different plants and animals which occupy the area increases, until a stable and balanced community of animals and plants is reached. This is known as a **climax community**.

COMBUSTION

Combustion (burning) is a chemical reaction which involves a substance reacting with oxygen:

1. Some metals will burn in air to produce metal oxides; how easily they do this depends on how reactive the metal is.

    magnesium + oxygen → magnesium oxide

2. Fuels are substances that give out a lot of energy when they burn; their reactions are strongly **exothermic**. The fuels react with the oxygen in the air. The **fossil fuels** we use (coal, gas and oil) contain carbon and hydrogen and are referred to as **hydrocarbons**. The products of combustion are carbon dioxide and water.

- *Coal* is mainly carbon:

    carbon + oxygen → carbon dioxide
    $C + O_2 \rightarrow CO_2$

- *Natural gas* is methane:

    methane + oxygen → carbon dioxide + water
    $CH_4 + 2O_2 \rightarrow CO_2 + 2H_2O$

▶ COMBUSTION IN ACTION

Natural gas is often considered to be a very suitable fuel for greenhouses. This is because it not only produces heat but also *carbon dioxide*, which the plants can use, as well as water, to keep the atmosphere humid.

However, if a *limited* amount of air is present, then *carbon monoxide* (CO) will be produced. This gas is poisonous because it latches on to Haemoglobin in the blood, forming carboxy-haemoglobin, and preventing the blood from carrying oxygen around the body. For this reason it is important to keep a room well ventilated when coal or gas is being burned.

◀ Reactivity series ▶

COMMUNITY

This term describes a group of animals and plants which live in a certain habitat, and interact with each other forming the ecosystem. For example, in a woodland there will be trees, shrubs and grass, being eaten by birds, worms, insects and mammals. All the organisms in the woodland make up a 'community'. ◀ Ecosystem, food chain, food web ▶

COMMUTATOR

◀ Motor ▶

COMPARATOR

A comparator is the *control mechanism* in a feedback system which initiates any corrective measures. For example, in an oven the *thermostat* will switch on the circuit to the heating elements. ◀ Control systems ▶

COMPETITION

When two organisms both require the same resource, such as food, space, light, and water, they can be said to be *competing* with each other. Sometimes competition occurs between members of the same species, and sometimes between members of a different species.

Figure C.9 shows how the population of a predator increases and decreases as the population of the prey goes up and down.

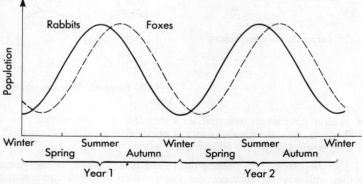

Fig C.9 Competition

COMPOSITE MATERIALS

Composite materials are those which are **mixtures**. Concrete is a composite material since it is a mixture of cement, sand and gravel.

Fillers are often added to a **polymer** to give it extra strength. For example, glass fibres are added to a **thermosetting plastic** to give it enough strength to be used to make the glass fibre hulls of boats. Military aircraft have wings made up of a composite material containing carbon fibres which are much lighter than metal. Many modern cars are made using a glass reinforced plastic for the bonnet lids and boot lids.

COMPOUNDS

Compounds are substances which contain more than one type of **atom**. These atoms are chemically joined in a compound. The particles in a compound might be **molecules** or **ions**. A compound can be chemically split into simpler substances.

For example:

- *water* is a compound, it consists of water molecules. Each water molecule contains two atoms of hydrogen and one atom of oxygen (Fig C.10a)). Water can be split by **electrolysis** into hydrogen and oxygen.

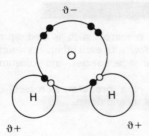

Fig C.10a) Compounds: the water molecule

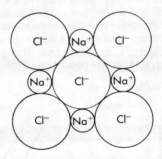

Fig C.10b) Compounds: the sodium chloride lattice

- *sodium chloride* is a compound, it consists of sodium and chloride ions. There is one sodium ion for every chloride ion (Fig C.10b)). It can be split by electrolysis into sodium and chlorine.

In Fig C.10c) we list the melting points and boiling points of compounds, together with whether they conduct when molten. You will notice that all metal compounds conduct electricity when molten.

CONDUCTION

Compound	Relative molecular mass	Conduct when molten	M.p.°C	B.p.°C
ammonia	17	no	− 78	− 34
carbon dioxide	44	no	− 111	− 78
calcium carbonate	100		decomposes when heated	
calcium oxide	56	yes	2600	2850
copper(II) chloride	135	yes	620	990
copper(II) sulphate	160		decomposes when heated	
glucose	180	no	146	decomposes
hydrogen chloride	36.5	no	− 114	− 85
lead(II) chloride	278	yes	501	950
silicon dioxide	60	no	1610	2230
sodium chloride	58.5	yes	801	1413
water	18	no	0	100

Fig C.10 c)

CONCENTRATION

◀ Molar solutions ▶

CONDENSATION

Condensation is the liquid formed as a result of cooling a gas; for example in a warm room there may be a lot of water vapour which, when it reaches the cold glass of a window, condenses out to form water droplets.

CONDITIONED REFLEX

◀ Reflex arc ▶

CONDUCTION

This term describes how heat is transferred in a solid. The solid itself does *not* move, as the process of conduction depends on energy being passed from one **molecule** to another. For example, metals are good conductors of heat but non-metals are poor conductors.

A group of similar-sized solid rods, heated as shown, can illustrate the idea of rates of conduction. The rods have a matchstick sealed to one end with wax. If they are equally heated, the matchsticks fall at different times. Heat is flowing along the rods without the material itself moving. This is an example of conduction. Generally, metals are good conductors, and non-metallic solids tend to conduct badly, and are called **insulators**.

CONSUMERS

Conductor	Insulator
Copper	Wood
Iron	Sulphur
Aluminium	Polythene
Carbon	Rubber
Sea-water	Paraffin
Sulphuric acid	Propanone

The mechanism of conduction relies on the passing of kinetic energy form one vibrating particle to another, so a good conductor will have strong linkages between particles. Since the inter-particle linkages between particles are not as strong in liquids as in solids, and are almost non-existent in gases, liquids and especially gases are very poor conductors. ◄ Insulation ►

CONSUMERS

Consumers are the animals in a **food chain**. They obtain their energy from the green plants which are the **producers** in a food chain. Consumers can be **herbivores** which feed directly on green plants; **carnivores**, which feed on other animals; or **omnivores** which feed on plants and animals.

CONTROL SYSTEMS

All systems, whether biological, physical or chemical, need to be kept in control and stabilised. If a biological system in your body becomes unstable, you become ill and may die. Mechanical, chemical and electronic systems will stop working if they go out of control.

You use control systems in everyday life, such as a *central heating thermostat* to control the temperature of a room, a *ballcock* (Fig C.11a)) to control the level of the water in the toilet cistern, and a *carburettor* to control the flow of petrol and air to a car engine.

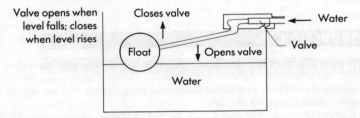

Fig C.11a) Automatic water control – a ballcock

You may have used electronic control devices such as a **light-dependent resistor**, which will turn a switch on and off depending on the light intensity.

Three components are needed in a control system, and what is being controlled is described as the *controlled variable* (Fig C.11b).

CORNEA   63

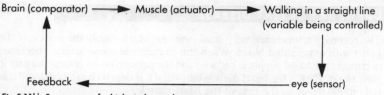

Fig C.11b) Components of a biological control system;

The three components are:

- the **receptor** or *sensor* which detects change,
- the **comparator** or control mechanism which initiates the corrective measures.
- the **effector** or **actuator** which brings about the corrective measures.

◀ Feedback systems; thermostats ▶

CONVECTION

This term describes how heat is transferred in a fluid. Hotter, less dense, liquids rise up; colder, denser fluids sink down to take the place of the hotter fluids. A *convection current* is produced as a result of the continuous movement of the fluids.

You may have studied convection currents by placing a crystal of potassium permanganate in a beaker of water. As you gently heat the water containing the crystal, the liquid becomes less dense and rises. Colder, denser water sinks to take its place (Fig C.12).

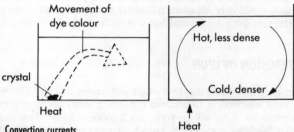

Fig C.12 Convection currents

CORE

The core of the **Earth** is mostly a mixture of nickel and iron. The *inner core* is solid iron-nickel alloy, whereas the *outer core* is liquid iron sulphide. The core is about 7000 km in diameter.

CORNEA

◀ Eye ▶

CORONARY ARTERY DISEASE

The coronary arteries are very small **arteries** which supply the muscle of the **heart** with oxygenated blood. When the diameter of these arteries becomes narrower, the blood supply is reduced and the person feels a cramp-like pain in their chest. Part of the heart muscle may die if it is deprived of oxygen and this puts more strain on the rest of the heart muscle.

Two factors contribute to coronary heart disease:
- cholesterol fat builds up on the inside of the coronary arteries;
- blood clots can stick to the cholesterol and stop the blood flowing.

Exercise can help to reduce the risk of coronary artery disease by improving the functioning of the heart. A balanced **diet**, which reduces the amount of fats, can prevent cholesterol building up, stopping smoking and limiting alcohol intake can also help reduce the risk of heart disease.

CORROSION

Corrosion is a chemical reaction. Corrosion of a metal will only take place if the metal is in contact with a solution containing **ions**. when the metal corrodes, it loses **electrons** to form positive ions:

Metal atom − electron(s) → metal ion

$M(s) - e^- \rightarrow M^+$

Metals corrode at different rates, depending on their position in the **reactivity series**; *magnesium* will corrode more quickly than *copper* because magnesium is higher in the reactivity series and has a greater tendency to form ions.

▶ CORROSION OF IRON

Iron will corrode (rust) when it is in contact with water and air. The water is acting as a weak **electrolyte** (a solution containing ions) because it contains dissolved substances. Iron will corrode much more quickly when in contact with sea water, because sea water is a much stronger electrolyte (it contains a larger amount of dissolved salts). This is a real problem for ships; also cars that are kept near the sea (seaside towns) tend to corrode more quickly than their counterparts inland.

The reaction of iron corrosion is as follows:

$Fe(s) - 3e^- \rightarrow Fe^{3+}(aq)$

Corrosion can be a greater problem with structures built of *more than one metal*. For example, if the steel plates of a ship's hull are rivetted together with *brass* rivets, then the steel will corrode much more quickly than if the rivets were made of *steel*. This is because:

COVALENT BONDING

- the sea water is acting as the electrolyte;
- a simple cell is set up between the iron (steel) and the copper (in the brass). The iron has a greater tendency to form ions than the copper so will corrode much more rapidly.

RATES OF CORROSION

The *rate of corrosion* of a metal therefore depends on:
- the position of the metal in the reactivity series;
- the concentration of the electrolyte with which the metal is in contact;
- the nature of any other metal with which it is in contact;
- the temperature of the metal (this is why car exhausts corrode quickly) – a higher temperature speeds up chemical reactions.

PREVENTING CORROSION

1. **Surface coating:** This is the simplest method of preventing metals from corroding, by either painting the metal or covering it with some other covering such as a *polymer layer* which is bonded to the surface. Some oil rigs in the North Sea are protected in this way.
2. **Sacrificial protection:** This is where one metal is in contact with a more reactive one which corrodes first, e.g. galvanizing.
3. **Electroplating:** This is a method of preventing corrosion by covering one metal with a less reactive metal.

COULOMB

A coulomb is a word used to describe a very large number of charged particles, 6.2×10^{18}. When an ammeter measures that 1 amp of current is flowing, this really means that in one second a coulomb of charge is passing that point. A household lamp uses about 2 amps of current, which means that 2 coulombs are used per second. The amount of current flowing is therefore charge divided by time.

An easier way to remember this is:

charge = current × time   or $Q = I \times t$

So if 10 amps (I) flow for 5 seconds (t) then 50 coulombs (Q) have passed through a point in the circuit.

COVALENT BONDING

Atoms can have stable arrangements if their outer **electron** shells are filled. Atoms join together, forming bonds, in order to achieve this. When this happens by the sharing of electrons, it is called *covalent bonding*.

COVALENT BONDING

IN ELEMENTS

Non-metal atoms will combine to form **molecules** by sharing electrons in their outer shells. The exception to this are the atoms in Group 0, which have stable electron arrangements already.

Two *hydrogen atoms* will join together to form a *hydrogen molecule*, by sharing their electrons. Each atom can then be considered to have a filled *electron shell* (2 electrons). The shared pair of electrons is called a *covalent bond* (Fig C.13a)) and can be shown as a line between the two atoms '–'. The molecule is represented as H_2, or H–H.

Fig C.13 Covalent bonding

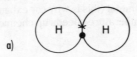

a)

An *oxygen molecule* is formed in a similar way but because each oxygen molecule has 6 electrons in its outer shell (electron configuration 2,6), it has two 'spaces' to be filled. It does this by each atom sharing *two* of its electrons. This forms a *double covalent bond* (Fig C.13b). The molecule is represented as O_2, or O = O.

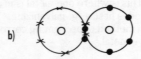

b)

Similarly, nitrogen atoms will pair up to form *nitrogen molecules* but this time by forming a *triple* covalent bond. Each covalent bond is a shared pair of electrons (Fig C.13c)). The nitrogen molecule is represented as N_2, or N≡N.

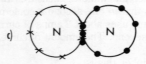

c)

IN COMPOUNDS

Non-metal atoms exist in the free state as **molecules** because in this way they can have *stable electron arrangements*. Non-metal atoms will combine with other non-metal atoms to form molecules. The resulting compounds are called *covalent compounds*, for example water H_2O (Fig C.13d)).

d) A water molecule

Some common covalent compounds

Figure C.14 shows the molecular structure of some common covalent compounds. Remember, *each line* represents a shared pair of electrons – a covalent bond.

Name	Formula	Structure
carbon dioxide	CO_2	$O=C=O$
ammonia	NH_3	$\begin{array}{c} H \\ | \\ N \\ / \ \backslash \\ H \quad H \end{array}$
sulphur dioxide	SO_2	$O=S=O$
methane	CH_4	$\begin{array}{c} H \\ | \\ H-C-H \\ | \\ H \end{array}$
ethane	C_2H_6	$\begin{array}{c} H \ \ H \\ | \ \ | \\ H-C-C-H \\ | \ \ | \\ H \ \ H \end{array}$
ethene	C_2H_4	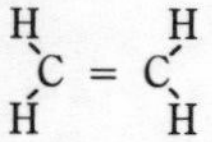 $\begin{array}{c} H \quad\quad H \\ \backslash \quad\ / \\ C=C \\ / \quad\ \backslash \\ H \quad\quad H \end{array}$
ethyl alcohol	C_2H_5OH	$\begin{array}{c} H \ \ H \\ | \ \ | \\ H-C-C-O-H \\ | \ \ | \\ H \ \ H \end{array}$

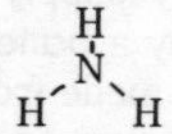

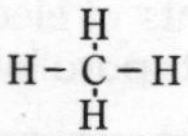

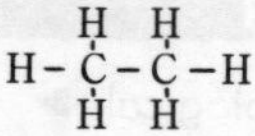

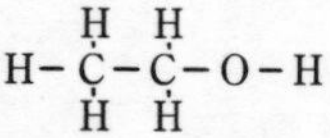

Fig C.14

Properties of covalently bonded molecules

The covalent bond between atoms in a molecule is strong; however the forces holding the molecules together are weak. This means that covalently bonded substances are often gases or liquids, or solids with relatively low melting points. These substances have low melting points, low boiling points, do not conduct electricity and do not usually dissolve in water.

CRACKING

Cracking is a term used in the petrochemical industry for breaking large hydrocarbon molecules down into smaller ones. For example, ethane (C_2H_6) is passed through a cracking tower with heated steam to produce the smaller ethene (C_2H_4) molecule.

CRUST

The Earth's crust is a hard outer layer about 70 km thick around the **Earth**. It consists mostly of **metamorphic rocks** and some **sedimentary** and **igneous rocks**. Many of the rocks contain minerals, which exist either as *pure elements*, such as gold, or as *mineral ores*, such as malachite (copper carbonate).

The Earth's crust contains large plates of rock, forming the continents, which float on the molten mantle. Mountain building and **volcanoes** can occur when two of these plates meet, whereas **earthquakes** can occur when the plates slide past each other.

CURRENT ELECTRICITY

An electric current is a flow of charged particles around a circuit. All circuits require an energy supply to push the particles around the circuit. This is usually provided by a **battery**, a laboratory power pack, or mains electricity. All circuits should be made from good **conductors**, such as metals, as these allow the charged particles to pass through easily. Some substances are poor conductors or **insulators**, such as plastic and glass, and these do not allow significant amounts of electricity to pass through.

◀ Series circuit, parallel circuit ▶

CYTOPLASM

◀ Cell – biological ▶

DARWIN

◀ Natural selection ▶

DAY

A day is the time taken for the **Earth** to spin on its own axis, about 24 hours.
◀ Seasons, year ▶

DEAMINATION

Deamination is the process of breaking down excess **amino acids** in the liver. **Urea** is produced as a waste product and is removed from the blood by the **kidney**, and passed out of the body in the **urine**.

DECOMPOSERS

Decomposers are fungi and bacteria which obtain their energy by breaking down the bodies of dead animals and plants and releasing nutrients such as nitrogen into the soil. ◀ Food chain, nitrogen cycle ▶

DEFICIENCIES

Many people in under-developed countries suffer from a lack of one or more of the different types of food from their daily diet. For example, a lack of protein causes a disease called **kwashiorkor** and children are unable to grow and develop properly. A lack of vitamin A causes 250,000 children to go blind every year, and many more children suffer severe eye problems. A lack of **iodine** in the diet causes many children in less developed countries to suffer mental retardation. ◀ Balanced diet ▶

DENSITY

The density of a substance is the mass of the substance divided by its volume; units are kg/m^3 or g/cm^3.

$$\text{Density} = \frac{\text{Mass}}{\text{Volume}} \quad \text{or} \quad D = \frac{M}{V}$$

Each substance has a different density and this fact can help to identify a substance. The density of a substance is *high* if a large mass occupies a small volume; for example: steel and lead. If the density of a substance is *low*, it has a small mass occupying a large volume; for example: gases.

DEOXYRIBONUCLEIC ACID (DNA)

◀ Chromosomes ▶

DETERGENTS

The job of a detergent is to help water clean materials by washing away grease and dirt. Water is a good solvent, particularly for ionic substances, but it is not good at dissolving greasy substances. The detergent molecules help by breaking up the grease into smaller globules which can be carried away by the water.

How detergents work

Detergent molecules consist of long carbon chains which have two different parts. One part of the molecule is water-loving (*hydrophilic*) because it carries an electric charge (ionic). The other part of the molecule is water-hating (*hydrophobic*) because of the long covalently bonded chain. To help explain the process we can represent these molecules as pin shapes as shown in Figure D.1.

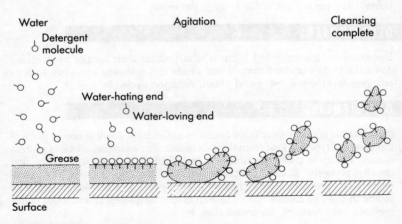

Fig D.1 Detergents

The water-hating (or grease-loving) part of the molecule buries itself in the grease while the water-loving part stays in the water. During agitation the water can get between the grease and the surface to be cleaned. The grease forms globules that are kept apart by the charges on the detergent ('like' charges repel). The detergent and grease form an emulsion in the water.

Detergents in action

There are two types of detergents. *Soap detergents* are made from animal or vegetable fats and alkalis. Sodium stearate (see Fig D.2) is a typical soap. In water, the sodium ion floats free, leaving one end of the molecule negatively charged.

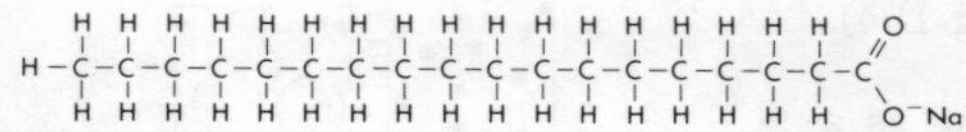

A soap molecule

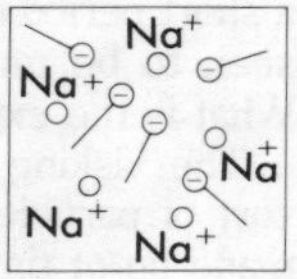

Soap molecules in water

Fig D.2 A typical soap molecule (sodium stearate)

Soapless detergents are made using chemicals from oil and acids. A typical soapless detergent is shown in Figure D.3. Again in water, the sodium ion is separated from the molecule, leaving on end negatively charged.

◄ Hard water ►

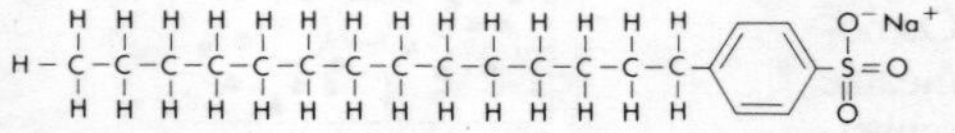

A detergent molecule

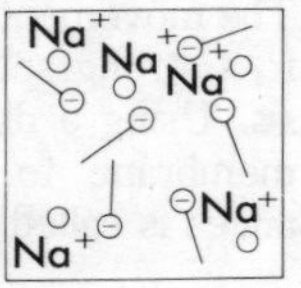

Detergent molecules in water

Fig D.3 A typical soapless detergent molecule

DIALYSIS

Using a **differentially permeable membrane** (DPM) to separate substances. Kidney machines use flat tubes of cellophane as DPMs.

DIAPHRAGM

◄ Breathing ►

DIET

◄ Balanced diet ►

DIFFERENTIALLY PERMEABLE MEMBRANES

A concentrated sugar solution is placed in a bag made of **visking tubing** (a material like cellophane). The bag and its contents are then placed in a beaker of dilute sugar solution. After a short period of time the bag will be seen to be much bigger (Fig D.4). What is the explanation?

The visking tubing is acting as a sort of particle sieve. The tubing is made up of tiny holes or pores, just big enough to let the small water particles (which are moving) pass through, but *not* the large sugar particles.

Materials such as visking tubing are called differentially permeable membranes. If allowed to continue, water would pass through the tubing walls until the concentration of the solutions inside and out is the same. The movement of water from a weak to a strong solution is called **Osmosis**. Using a differentially permeable membrane to separate the substances is called **dialysis**.

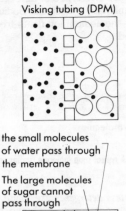

the small molecules of water pass through the membrane

The large molecules of sugar cannot pass through

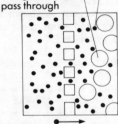

Fig D.4 Differentially permeable membrane

 DPMs IN ACTION

In cell membranes

Cell membranes act as DPMs allowing water to enter cells. Water passes into root hairs (special cells on the tips of roots) by osmosis. The cells contain a solution of sugars, salts and other solutes; water can enter through the cell membrane because the solution inside the cell is *more concentrated* than outside.

In the kidneys

The **kidneys** do an important job. They get rid of the waste products from our bodies as well as maintaining the balance of blood sugars, salts and water in the body. The kidney machine which can take over when some people's kidneys fail due to disease uses DPMs (flat tubes of cellophane). Harmful substances are removed from the blood by osmosis.

In food preservation

Some methods of food preservation work because of DPMs and osmosis. The bacteria which cause food to decay are single-celled organisms and their cell membranes act as DPMs. Fruit can be *preserved* by placing it in a concentrated sugar solution (for example, jams and preserves). Any bacteria that *do* get into the jam are killed because they become *dehydrated*. The solution outside the cell is more concentrated than the solution inside, so water moves *out* of the cell. Food preservation by salting can be explained in the same way.

DIFFUSION

Diffusion means mixing; gases, liquids and even solids can mix together or diffuse if left alone (even without anyone stirring them!)

TYPES OF DIFFUSION

Diffusion in gases

When the top is removed from a bottle containing *ammonia* you can smell the ammonia (which is a gas) even if you are some distance away. The ammonia has mixed with the air and spread out.

When the equipment in Figure D.5 is set up and left for a short time, a white ring appears in the tube. This white substance is *ammonium chloride* which is formed when the gas ammonia meets the gas hydrogen chloride. This could only happen if the gas particles had **kinetic** energy and were able to spread out and mix. Notice the white ring where the gases meet is *not* in the centre. Which gas spreads out the fastest? Does this tell you which has the lightest particles?

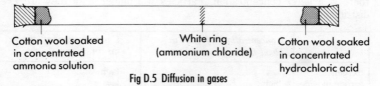

Fig D.5 Diffusion in gases

Diffusion in liquids

Diffusion can also occur in liquids. This can be shown by placing a crystal of *potassium permanganate* in a beaker of water. The diffusion takes place more slowly than between gases because the particles have less energy. They are moving more slowly and are closer together (Fig D.6).

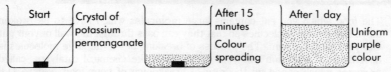

Fig D.6 Diffusion in liquids

Diffusion in solids

Diffusion can take place in solids although this takes place even more slowly. This can be shown by placing a *coloured crystal* in some gelatine; after a day or two the colour has spread through the gelatine (Fig D.7).

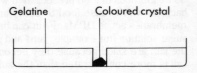

Fig D.7 Diffusion in solids

The only way to explain all these results is to assume that substances are made of *particles* and that these particles have kinetic energy (are moving).

▶ DIFFUSION IN ACTION

Atmospheric pollution

Diffusion can be a nuisance: it is because of diffusion that pollutant gases can spread throughout the atmosphere for example, from car exhausts, power stations and aerosols.

In the lungs

Gas exchange in the alveoli of the lungs takes place by diffusion. Diffusion of particles takes place from where there is a *higher* concentration to where there is a *lower* concentration. Particles will diffuse until they are evenly distributed:

	Concentration of gas in blood flowing to alveoli	Concentration of gas in air in alveoli
oxygen	low	high
carbon dioxide	high	low

◀ Breathing ▶

DIGESTION

The food which you eat contains large **molecules** which have to be broken down into smaller molecules so that they can pass through the wall of your gut into your blood stream. The process of breaking down the large molecules is described as 'digestion'. In your body you make chemical catalysts, called **enzymes** which speed up the rate of breakdown of your food. Figure D.8 shows the human digestive system in detail.

DIGESTION

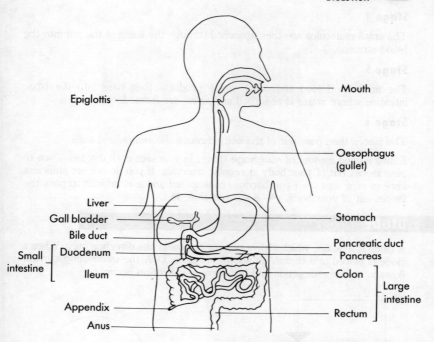

Fig D.8 Digestion: the human digestive system

STAGES OF DIGESTION

Stage 1

The process of digestion starts in your mouth when your teeth crush and chew the food which is mixed with saliva. The saliva contains an enzyme called **amylase** which starts to breakdown or digest the large molecules of starch.

Stage 2

The food is then swallowed and goes to your stomach for about 3 or 4 hours. The stomach lining secretes a dilute acid to create an acidic environment so that the enzyme *pepsin* can break down the large **protein** molecules into smaller molecules.

Stage 3

The partly digested food is then passed into the next part of the gut, the small intestine, where enzymes from the *pancreas* continue the process of digestion. The *starch* is eventually converted into small molecules of glucose; the *protein* is broken down into amino acids, and the *fats* are broken down into fatty acids.

Stage 4

The small molecules are then absorbed through the lining of the gut into the blood stream.

Stage 5

The undigested food and other waste products then pass into the large intestine where water is absorbed and *faeces* are formed.

Stage 6

The faeces then pass out of the body through the *rectum* and anus.

Eating a good amount of **roughage** (fibre) in your diet each day helps you to pass faeces out of your body at regular intervals. If you do not eat sufficient fibre in your diet then you become constipated and it is difficult to pass the faeces out of your body. ◄ Balanced diet ►

DIODE

A diode is a device which allows current to flow in one direction only. When a diode is placed in a circuit, as shown in Figure D.9, the **alternating current** flows in only one direction and half-wave **rectification** occurs.

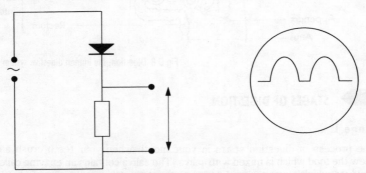

Fig D.9 Diode: the pattern produced on a CRO when a diode is placed in the circuit

DIPLOID

◄ Fertilisation ►

DIRECT CURRENT

Direct current is electric current which flows in one direction only, as produced by electric cells (**batteries**), and **dynamos**.

DISCONTINUOUS VARIATION

◄ Variation ►

DISTILLATION

Distillation is a technique for separating a volatile liquid (one which has a low boiling point) from a less volatile liquid. For example, if a mixture of alcohol and water is heated, the alcohol evaporates first because it has a lower boiling point than water. If the vapour at the top of the heated liquid is passed through a condenser (such as a *Leibig condenser* – a tube around which is flowing cold water) then the vapour is condensed back into liquid alcohol and can be collected (Fig D.10). The trick is to maintain the temperature of the heated mixture at just the right level so that only the alcohol boils and not the water. ◄ Fractional distillation ►

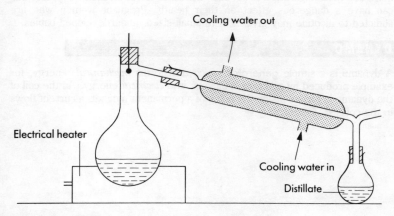

Fig D.10 Simple distillation

DOMINANT GENE

A dominant gene is the **gene** which is expressed in the appearance of an individual when both dominant genes are present, or when the dominant and **recessive** genes are present. For example, if a person has the dominant gene for brown eyes and the recessive gene for blue eyes, then the person appears to have brown eyes. ◄ Homozygous, heterozygous ►

DOUBLE GLAZING

◄ Insulation ►

DOUBLE INSULATION

Some electrical appliances have a plastic, insulated casing, so that there is no chance of someone getting an electric shock when they touch the appliance if it is faulty.

DRUGS

Drugs are chemical substances which either affect your body cells, or affect viruses and bacteria inside your body. Drugs are usually taken for medical reasons, but some drugs such as heroin are taken to produce a pleasurable sensation in the person. This can lead to drug addiction and death.

Certain drugs are *prescribed* by a doctor to help a person maintain good health. For example *insulin* is given to someone who has diabetes and is unable to control the amount of sugar in their blood. Some drugs are widely available but do *not* require a prescription. For example nicotine is an addictive drug found in cigarettes. People who are addicted to nicotine have to smoke more and more cigarettes to produce the same effect on their body, and this can have a dangerous effect on their health. Pregnant women who are addicted to nicotine may give birth to small sized, underdeveloped babies.

DYNAMO

A dynamo is a simple **generator** which transforms *mechanical* energy, for example produced by pedalling a bicycle, into *electrical* energy. As the coil of the dynamo is turned inside the poles of a permanent magnet, a current flows in the coil. ◄ Generator ►

EAR

The *ear drum* detects the compressions and rarefactions (changes in density) of the air which are caused when **sound waves** are produced from a vibrating source. The vibrations of the ear drum are passed through the three small bones or **ossicles** in the middle ear. The fluid in the *cochlea* or inner ear then vibrates and impulses are passed via the auditory nerve to the brain (Fig E.1).

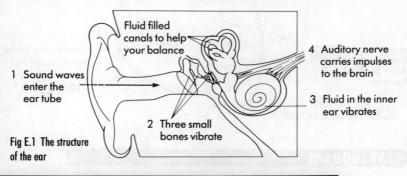

Fig E.1 The structure of the ear

EARTH

The Earth is a spherical planet of about 12,757 km diameter. It takes 365.25 days to orbit the Sun at a distance of about 150,000,000 km from the Sun. (Fig E.2).

The diagram is not to scale

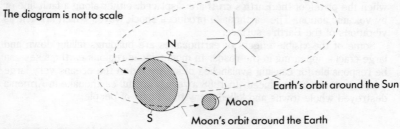

Fig E.2 The Earth's orbit around the Sun

EARTHQUAKE

The Earth is surrounded by an **atmosphere** about 400 km deep. The outer, solid part of the Earth is the **crust**, about 70 km deep, consisting of large sections of rock called *plates*. These plates are moved very slowly by the currents which flow in the *mantle* (a thick layer of very hot rocks). Where the plates move *against* each other, pressure builds up and may cause cracks (faults) to appear on the surface, and violent shaking movements (**earthquakes**). The crust contains **igneous, metamorphic** and **sedimentary rocks**.

The crust covers the mantle, a thick layer of very hot rocks, that in turn surround the core, which is made of very hot dense rocks under extreme pressure (Fig E.3).

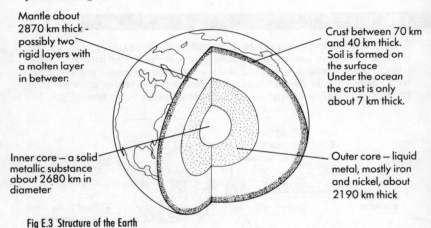

Fig E.3 Structure of the Earth

EARTHQUAKE

Earthquakes and **volcanoes** mostly occur in well-defined zones, usually at the boundaries of the different sections of the Earth's crust, for example around the coast of the Pacific Ocean. Other main earthquake areas are in a zone stretching from the Mediterranean through the Middle East, to the Himalayas, Indonesia and South China (Fig E.4). Earthquakes are caused when the plates of the Earth's **crust** are displaced, either along a fault line or by volcanic action. The earthquake produces shock waves which are felt as vibrations of the Earth's surface.

Some of the visible effects of earthquakes are buildings falling down and large cracks appearing in the roads. In mountainous regions earthquakes can be responsible for causing avalanches of snow, and in the oceans very large tidal waves can be produced. In 1988 a very powerful earthquake in Armenia destroyed whole towns and killed many thousands of people.

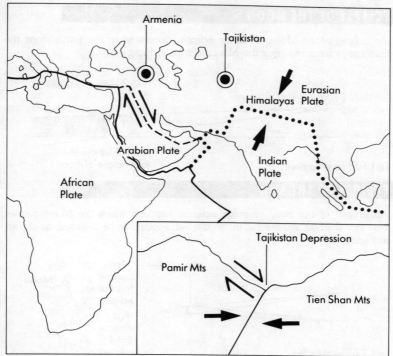

Fig E.4 Earthquake zones

EARTH WIRE

In a **three-pin plug** the *earth wire* is coloured green and yellow, and is connected to the earth pin at the top of the plug (Fig E.5). The purpose of the earth wire is to make sure that electricity flows to earth, if for any reason the appliance becomes faulty. This may happen if the live wire touches part of the metal casing of the appliance. If the earth wire was *not* connected, then electricity would flow to a person who touched the metal casing of the appliance.

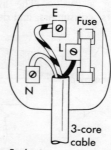

Fig E.5 Earth wire

ECLIPSE OF THE MOON

An eclipse of the Moon, or **lunar eclipse**, occurs when the Earth stops the Sun's rays from reaching the Moon, shown in Figure E.6.

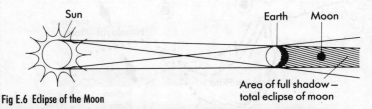

Fig E.6 Eclipse of the Moon

ECLIPSE OF THE SUN

An eclipse of the Sun, or **solar eclipse**, happens when the Moon passes between the Sun and the Earth, so the Sun appears to be covered, as shown in Figure E.7.

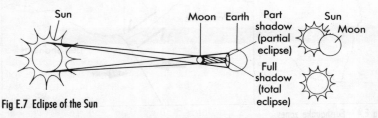

Fig E.7 Eclipse of the Sun

ECOSYSTEM

An ecosystem is all the living (**biotic**) and non-living (**abiotic**) factors in a specific area such as woodland, a pond or lake, a hedgerow, a sand dune, a beach, or a garden. The living components of an ecosystem are the **producers**, **consumers**, and **decomposers** (Fig E.8).

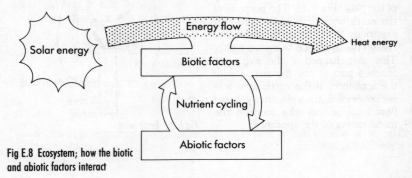

Fig E.8 Ecosystem; how the biotic and abiotic factors interact

EFFECTOR

An effector is usually a muscle or gland which responds to a stimulus. For example, when you touch a hot object it is the muscles of your upper arm which contract to move your hand away from the heat. ◀ Reflex arc, receptor ▶

ELECTRIC CELL

◀ Cell − electrical ▶

ELECTRICITY

◀ Current electricity, direct current, alternating current, series circuit, parallel circuit, generators ▶

ELECTRODE

An electrode is a positive or negative terminal in an electrolysis cell or an electrical cell (dry cell, battery, etc.). The negative electrode is called the **cathode**; the positive electrode is called the **anode**. ◀ Electrolysis ▶

ELECTROLYSIS

If electricity is passed into a *solution* or *molten substance* containing **ions** (both states where the ions are free to move) then the positive ions will be attracted to the negative electrode, and the negative ions will be attracted to the positive electrode. (*Electrodes* are rods which carry the current to the liquid and where ions are converted into **atoms**.) The end result is that electrolysis breaks down compounds containing ions into their **elements**.

An electrolysis cell always contains:

- a positive electrode, or *anode*;
- a negative electrode, or *cathode*;
- an electrolyte (a liquid containing ions, for example an aqueous solution of an ionic substance or a molten ionic substance).

The anode and cathode are connected to an electrical power source. The electricity is conducted through the liquid electrolyte by the ions themselves moving.

- The positive ions are attracted to the *cathode*.
- The negative ions are attracted to the *anode*.

At the cathode, the positive ions (metal or hydrogen) gain an electron and become *atoms*:

$$M^+ + e^- \rightarrow M$$

The metal is deposited as a layer on the cathode, any hydrogen is released as molecules of the gas.

ELECTROLYSIS

At the anode, the negative ions lose electrons and become atoms (which may combine to form molecules):

$$X^- - e^- \rightarrow X$$

The reactions occurring in a typical electrolysis are shown in Figure E.9, the electrolysis of molten sodium chloride.

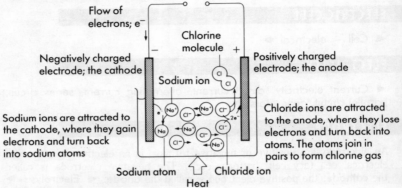

Fig E.9 Electrolysis of molten sodium chloride

EXAMPLES OF ELECTROLYSIS

Aluminium extraction

Aluminium is extracted from its ore by electrolysis. The ore *bauxite* (mainly aluminium oxide) is first concentrated by removing the impurities. The concentrate (alumina) is then dissolved in molten cryolite, a solution which provides free-moving aluminium ions – aluminium oxide has a melting point above 2000°C, so melting the oxide to provide free-moving aluminium ions is not practical. The anodes and cathodes are made of carbon; aluminium when it is formed is molten and is tapped off from the bottom of the cell (Fig E.10).

The following reactions are occuring in the electrolysis cell:

At the cathode: $Al^{3+} + 3e^- \rightarrow Al$

At the anode: $2O^- - 4e^- \rightarrow O_2$

All metals above aluminium in the reactivity series are normally extracted by electrolysis because they are too reactive to be reduced by carbon. Sometimes metals lower down such as zinc are extracted by electrolysis.

Fig E.10 The electrolysis of alumina to extract aluminium

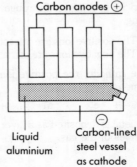

ELECTROMAGNETIC WAVES

ELECTROMAGNETIC RELAY

An application of the principle of **electromagnetism** is the *electro-magnetic relay*. The relay is a simple switch operated by an electromagnet in which a small input current controls a larger output current (Fig E.11). Stages 1 to 4 below describes how it works:

1. The input current causes the electromagnet to become magnetised.
2. The electromagnet attracts a soft iron armature which closes the contacts and causes a greater current to flow through the output circuit.
3. The output circuit controls a device such as a motor.
4. When the input current stops, then the output current is switched off and the motor stops.

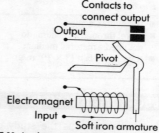

Fig E.11 An electromagnetic relay

ELECTROMAGNETIC WAVES

These are a group of transverse waves which have electric *and* magnetic properties. They are all produced by changing magnetic fields and electric fields, and travel at the very high speed of 300,000,000 metres per second.

```
long wavelength ◄─────────────────────────────── short wavelength
radio   micro   infra-red   visible   ultra-violet   X-rays   gamma   rays
low frequency ───────────────────────────────► high frequency
```

Fig E.12

Figure E.12 shows the position, relative wavelength and frequency of the different electromagnetic waves.

Type of wave	Uses	Source
radio wave: long wave medium wave short wave	radio communication	radio transmitters
VHF UHF micro waves	stereo radio television satellite communication, radar, microwave ovens	electronic circuits
infra-red	electric fires, ovens	any hot object
visible light	electric lights	very hot objects
ultra-violet	suntanning	extremely hot objects glowing gases
X-rays	used in hospitals to photograph bones	X-ray tubes
gamma rays	used to irradiate food to kill germs; can penetrate very dense metal	radioactive metals

ELECTROMAGNETISM

- Electromagnetic and mechanical waves.

The chart below summarises the main points of difference between *electromagnetic waves* such as radio waves, and mechanical waves, such as sound waves.

Electromagnetic	Mechanical
■ transverse waves ■ travel through a vacuum, do not need a material medium ■ travel very fast (3×10^8 m/s)	■ longitudinal waves ■ need a material such as air in which to travel ■ much slower speed (300 m/s approx)

ELECTROMAGNETISM

Electromagnetism is the study of the relationship between electricity and magnetism. A magnetic field is produced around a coil of wire whenever an electric current flows through the wire (Fig E.13). Plotting compasses or iron filings can be used to show this magnetic effect. These magnetic fields are fairly weak, and the effect can be increased by using a coil of wire called a *solenoid*. When the wire is coiled around a soft iron bar (called a *core*), the magnetic effect is more powerful, depending on the size of the current and the number of turns on the coil. Figure E.14 shows the current flowing in a clockwise direction around the X end of the core. This end becomes the south pole. When the current flow is anti-clockwise, then X becomes a N pole. So the polarity changes by reversing the current direction.

Fig E.13

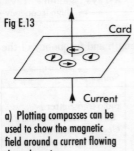

a) Plotting compasses can be used to show the magnetic field around a current flowing through a wire

b) The pattern produced for a single wire carrying current. The current is flowing upwards out of the page

c) The pattern produced for a single wire carrying current. The current is flowing downwards into the page

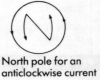

a) The current flows in a clockwise direction around the X end of the core

b) The current flows in an anti-clockwise direction around the X end of the core

Fig E.14

ELECTROMOTIVE FORCE

Electromotive force or E.M.F. is the force which drives an electric current around a circuit. The force comes from an **electric cell** (battery) or from a **generator**. E.M.F. is measured in **volts**.

ELECTRON

An electron is one of three **particles** found in an atom (the others being **proton** and **neutron** which are found in the **nucleus**). Electrons have very little mass (about 1/2000 that of a proton) and carry a negative charge. They orbit the nucleus of an atom in shells or orbits which take up most of the space of the atom. Each shell or orbit can only take a certain number of electrons (Fig E.15).

Chemical reactions between atoms are a result of interactions between the electrons in an atom, so chemical activity depends on the numbers of electrons in the outermost shells of an atom. There are always the same number of *electrons* in an atom as it has *protons*.

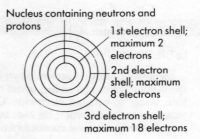

Fig E.15 Electron shells

ELECTROPLATING

Electroplating is a method of preventing corrosion by covering the surface of a metal to be protected with a thin layer of another metal which does *not* corrode. This is done by placing the metal to be plated in an **electrolysis** cell connected as the cathode.

Examples of electroplating to prevent corrosion are 'tin cans' (steel coated with a thin layer of tin), and chromium-plated bumpers on cars. Electroplating can also be used to produce cheaper products of precious metals, such as silver and gold (hence the term gold-plated or silver-plated). The main bulk of the object may be steel, coated with a thin layer of gold or silver giving the appearance of being made of solid gold or silver. Electroplating also ensures a thin, even coat of the metal.

ELEMENT

All substances can be classified according to the types of **particles** they contain and how these particles are joined (chemically bound or not). Substances can be classified as either *Element, Compound* or *Mixture*.

ELEMENT

Elements are substances that contain only *one* type of atom. Their particles may be single atoms or molecules. For example, hydrogen is an element; hydrogen gas contains hydrogen molecules. Copper is also an element; copper contains copper atoms (Fig E.16).

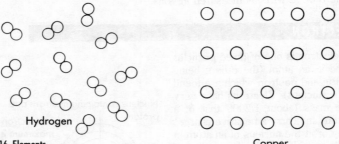

Fig E.16 Elements

Elements cannot be broken down into simpler chemical substances. Here the melting points and boiling points of the elements are listed together with their ability to conduct electricity. Only two elements are liquids at room temperature and pressure; they are bromine and mercury. You will also notice that all metals are good conductors of electricity:

Element	Relative atomic mass	Conductor of electricity	M.p.°C	B.p.°C
aluminium	27	yes	660	2450
argon	40	no	−189	−186
bromine	80	no	− 7	− 58
calcium	40	yes	845	1490
carbon	12			
diamond		no	3500	3900
graphite		yes	3500	3900
chlorine	35.5	no	−101	− 34
copper	64	yes	1083	2575
helium	4	no	−270	−269
hydrogen	1	no	−259	−253
iodine	127	no	114	184
iron	56	yes	1540	2900
lead	207	yes	330	1750
magnesium	24	yes	650	1100
mercury	210	yes	− 39	357
neon	20	no	−248	−246
nitrogen	14	no	−210	−196
oxygen	16	no	−218	−183
potassium	39	yes	63	760
silicon	28	yes	1410	2360
sodium	23	yes	98	880
sulphur	32	no	119	445
zinc	65	yes	419	907

EMBRYO

The embryo develops from the fertilised egg or **zygote**. It develops by cell division into a fully developed offspring, which in humans is called the *foetus*. In plants the embryo develops from a fertilised *ovule* (egg cell) and grows into a new plant.

ENDOTHERMIC

Energy is always involved in a chemical reaction; the substances reacting sometimes *take in* energy and sometimes *give out* energy.

Endothermic reactions are those in which energy in the form of heat is transferred from the surroundings, to the substances; in other words heat is taken in; often the *products* are cooler than the *reactants* were. Energy is needed for any chemical reaction to take place; first to break any bonds and then to reform new chemical bonds. If the energy required to break the chemical bonds is greater than the energy released when new bonds are formed, then the reaction is *endothermic* (Fig E.17).

Photosynthesis is an example of an endothermic reaction. When many substances dissolve in water the 'reaction' is often endothermic.

◀ Exothermic ▶

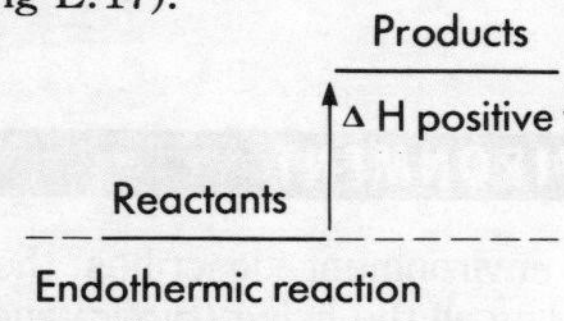

Products

↑ ΔH positive value

Reactants

Endothermic reaction

Fig E.17 Endothermic reaction: Energy level diagram

ENERGY

Energy is the ability to do work. The unit of energy is the **joule**. There are different forms of energy: **potential** (stored), **kinetic** (motion), **heat** (thermal), **light, electrical, chemical, nuclear**.

Energy can be transferred from one form to another, but the total amount of energy remains unchanged. The total quantity of energy entering a system is equal to the total quantity of energy leaving the system. For example, in a *hydro-electric power station* the kinetic energy of the water is converted into electrical energy and heat energy. The energy efficiency of the power station is therefore equal to the quantity of useful energy given out, divided by the total quantity of energy input.

$$\text{Energy efficiency} = \frac{\text{quantity of useful energy given out}}{\text{total quantity of energy input}}$$

▶ ENERGY FROM FOOD

The blood system carries the small molecules of glucose and amino acids, together with oxygen, to every cell in your body. A chemical process called **respiration** takes place in the cells to release the energy from the food:

food + oxygen → carbon dioxide + water + energy

The carbon dioxide and water are carried by the blood to your lungs and breathed out. The energy is used by your cells for the various functions of the cells. Muscle cells need energy for contraction, gut cells need energy for secretion and absorption.

You may have investigated the energy released from different foods by letting them burn under a measured volume of water, and measuring the change in temperature (Fig E.18). High energy foods are usually fats, such as butter and oils, and carbohydrates, such as potato, bread, and sugar.

◄ Respiration ►

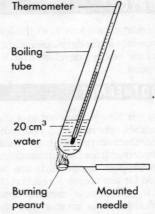

Fig E.18 Energy from food; finding out how much energy is released when a peanut is burned

ENVIRONMENT

The environment describes the surroundings where an organism lives, including all the living (**biotic**) and non-living (**abiotic**) factors which affect an **organism**. The living and non-living factors interact to produce a balanced **ecosystem**. The main environments are the sea, the rivers and lakes, and the land, which contain several **habitats**.

ENZYMES

Enzymes are biological catalysts. Many chemical reactions take place in living things (often in the cells). These reactions are controlled by *enzymes*.

Enzymes are **protein molecules**, consisting of long chains which can be folded and coiled into different shapes. Each enzyme has its own special shape; it is this shape which causes the enzyme to act as a catalyst. The reactant molecule(s) on which it acts fits the enzyme like a key in a lock (Fig E.19).

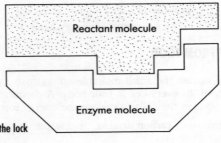

Fig E.19 Enzymes; the lock and key principle

PROPERTIES OF ENZYMES

- Enzymes are specific. Because the way an enzyme works depends on its shape it will only work for *one* molecule or reaction.
- Enzymes will only work within a small temperature range. For example, enzymes in the human body will only work around normal body temperature (37°C).
- Enzymes are very sensititve to pH changes and will only work within small pH ranges.
- Enzymes can catalyse many reactions. They can help large molecules break into smaller ones; help small molecules join to form larger ones, or help atoms rearrange within a molecule.

ENZYMES IN ACTION

In digestion

Enzymes take part in every part of the digestive process, helping to break large molecules into smaller ones so they may pass through the gut wall into the blood stream. For example, an enzyme in saliva called **salivary amylase** breaks down starch into smaller sugar molecules with ideal working conditions 37°C about pH 6–7; enzymes in the stomach which act on food, work at lower pH values (down to pH 2).

In brewing

In the brewing of beer, the first stage is mashing the barley. In this process natural enzymes in the barley convert the starch into fermentable sugars (later acted on by yeast). These enzymes work best at about 65°C and an acidity of about pH5–6.

In industry

Enzymes are becoming increasingly important in industry, for example in the manufacture of **biodegradable** dressings for wounds, in new food sources such as mycoprotein, and in biological washing powders. ◄ Digestion ►

EROSION

Erosion describes the way in which the Earth's surface is being worn away (eroded) by weathering agents such as wind, ice or frost, changes of temperature, chemical action, or the action of living organisms. The result of erosion is to form much smaller **particles** which are then carried by wind or water and eventually form soil. ◄ Sedimentary rocks ►

EVAPORATION

If a bowl of water is left on a windowsill the water will eventually disappear; a puddle of rain will also eventually disappear; we say the water has *evaporated*. The same is true of liquids other than water.

EVAPORATION

In a beaker of a liquid (for example water) the particles (water molecules) will have different energies, some fast-moving molecules will have enough energy to *escape* the surface of the water. This is evaporation. When the molecules escape the surface, they bump into air molecules and some might even travel back into the liquid. The more molecules that escape the surface and become vapour (gas form), the more chance that any newly escaped molecules will be knocked back into the water (Fig E.20).

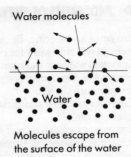

Molecules escape from the surface of the water

Fig E.20 Evaporation

We can increase the speed of evaporation in several ways:

- blowing across the surface of the water, removing the vapour molecules as they are formed (this is why blowing across a cup of tea will cool it down).
- heating, thereby giving more molecules the energy to escape. If enough energy is transferred to the water, *all* the molecules will be able to escape. We call this *boiling*; water boils at 100°C at sea level.
- by reducing the air pressure, allowing the molecules to escape more easily. This is why water will boil at a lower temperature on mountains (less air pressure) than at sea level.

▶ EVAPORATION IN ACTION

Pressure cookers

Pressure cookers cook food more quickly because they allow the water to boil at a higher temperature. The increased pressure prevents the high energy water molecules escaping the surface of the water.

Aerosols

Aerosol cans contain liquids under pressure that are normally gases at atmospheric pressure (these liquids have boiling points just below room temperature). Pressing the nozzle releases the pressure, allowing some of the liquid (called the *propellant*) to evaporate, carrying the substance to be sprayed with it. Much concern has been shown recently because some of these propellants called **chlorofluorocarbons** (CFCs) have been shown to damage the **ozone layer**.

Controlling body temperature

We cool down more quickly if our skin is wet. This is because our body heat evaporates the water molecules on our skin, taking the heat energy with them.

EVOLUTION

This describes the way in which different species of organisms have developed and changed over a very long period of time (Fig E.21).
◄ Natural selection, mutation ►

Recent ↑

Name	Skull	Fore limb	Hind limb	Teeth Top view	Teeth Side view	Height (cm)
Equus						150
Pliohippus						125
Meryo hippus						100
Meso hippus						60
Eohippus						28
Ancient	Hypothetical ancestor with five toes on each foot and monkey-like teeth					

Fig E.21 Evolution

EXCRETION

Excretion is the removal of waste products formed in the body by metabolic processes (Fig E.22). For example, the liver breaks down excess amino acids during the process of **deamination** and produces **urea**, a waste product. The urea is carried in the blood stream to the kidneys, where it is removed from the blood and passed out of the body in the urine. The skin also removes some urea during sweating. The lungs remove carbon dioxide, a waste product produced by the cells during **respiration**.

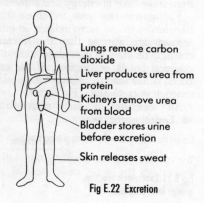

- Lungs remove carbon dioxide
- Liver produces urea from protein
- Kidneys remove urea from blood
- Bladder stores urine before excretion
- Skin releases sweat

Fig E.22 Excretion

EXERCISE

Your muscles need more energy when they are working harder during vigorous *exercise*. Your *heart rate* increases to pump blood carrying glucose more quickly to your cells, and your *rate of breathing* increases so that more oxygen is taken in to release the energy from the glucose. More carbon dioxide is produced, which is removed by the increased rate of breathing. People who are fit generally have a lower heart rate and therefore a lower pulse rate than people who are unfit, because exercise develops the heart muscle, just like any other muscle. Fit people and non-smokers get back to their resting pulse rate more quickly than unfit people and smokers (Fig E.23). Regular exercise, eating a good well balanced diet without too much fat, and not smoking, can reduce the risk of heart disease.

◀ Balanced diet ▶

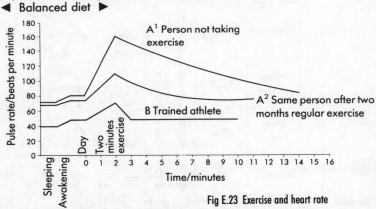

Fig E.23 Exercise and heart rate

EXOTHERMIC

Energy is always involved in a chemical reaction; the substances reacting sometimes take in energy and sometimes give out energy.

Exothermic reactions are those in which energy in the form of heat is transferred to the surroundings, in other words *heat is given out* – often the *products* are warmer than the *reactants* were. Energy is needed for any chemical reaction to take place, at first bonds have to be broken and then new chemical bonds are formed. If the energy required to break the chemical bonds is *less* than the energy released when new bonds are formed, then the reaction is *exothermic* (Fig E.24).

All combustion and neutralization reactions are exothermic.

◀ Endothermic ▶

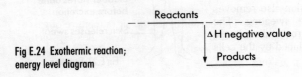

Fig E.24 Exothermic reaction; energy level diagram

EXPANSION

In general when matter is heated it *expands*; some substances expand more than others. This can be understood if we imagine all substances to be made of particles. Heating transfers energy to the substance increasing the **kinetic energy** of its particles (Fig E.25). A common mistake is to say that the particles themselves get bigger; this is not so, it is the *gaps* between the particles that increase.

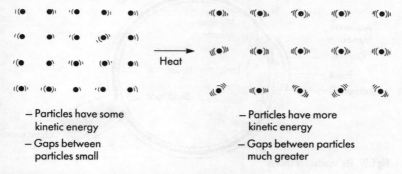

- Particles have some kinetic energy
- Gaps between particles small

- Particles have more kinetic energy
- Gaps between particles much greater

Fig E.25 Expansion; expansion is caused by increased kinetic energy of particles

EXPANSION IN ACTION

Bimetallic strip

This consists of two metals with different expansion rates (*coefficients of expansion*) stuck together. When heated, the strip bends. Bimetallic strips are often used in thermostats e.g. in an iron, as a switching device, as well as in flashing light bulbs (Fig E.26).

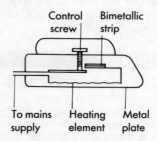

Fig E.26 Expansion; bimetallic strip in an iron

Railway lines

Gaps are left between rails on railways to allow for expansion. If no gaps were left, the rails may buckle in very hot weather.

Hot-air balloons

When gases are heated they expand and become less dense. This is why hot-air balloons rise into the air. The burner heats the air (a mixture of gases) which expands and becomes less dense, i.e. the gas molecules are further apart. Because the air inside the balloon is less dense than the colder air outside, the balloon rises into the air.

EYE

The *retina* of the human eye detects coloured light. On the retina are cells which are sensitive to red, green and blue light. When these cells are stimulated by light, an impulse is sent to the brain, via the *optic nerve*. The brain then forms images as a result of the impulses it receives. Figure E.27 shows the structure of the main parts of the eye.

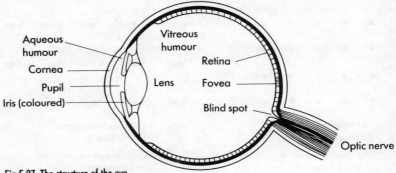

Fig E.27 The structure of the eye

Light is refracted, or bent, as it enters the eye through the transparent *cornea*. It is then refracted even more by the convex lens of the eye, and focused on the retina at the back of the eye. The lens is able to adjust to looking at objects which are close to or far away. This is known as *accommodation*.

FAULT

The pressure of the Earth causes the rocks in the crust to break or fault, as shown in Figure F.1a).

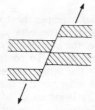

Fig F.1a) Fault

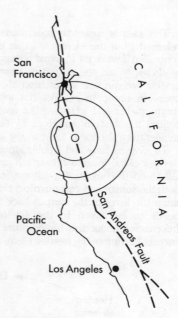

Fig F.1b) The San Andreas Fault

Earthquakes happen near large faults (e.g. the San Andreas Fault) as two plates in the Earth's crust move past each other, aided by lubrication of a small proportion of molten material (Fig F.1b)). Sudden fracturing releases energy, causing vertical and horizontal vibrations. Volcanic eruptions occur at weak places in the Earth's crust, as molten rock forces its way to the surface. In this area, lava runs from long fractures or *fissures* to form basalt plateaus.

FEEDBACK SYSTEMS

Feedback is the way in which a system controls itself. Some of the energy from the *output* of a system is passed back to the *input* of the system so that the input can be regulated. All systems that we want to control will consist of inputs and outputs (Fig F.2).

FEEDBACK SYSTEMS

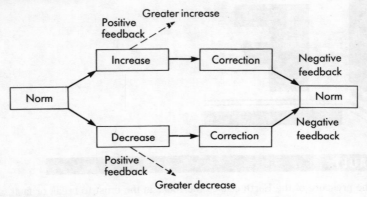

Fig F.2 Feedback mechanisms

This idea is easier to understand if we look at an example: heat energy released from the elements in an electric oven needs to be very carefully controlled if we want to avoid burning the food. Here, the input is the *amount of electricity* and the output is a *constant oven temperature*. To control this system we can use information about the output – the temperature of the oven, in order to control the input of electrical energy. By reading a thermometer and adjusting the amount of electricity being used, we could control the system ourselves by turning up the electrical supply when the temperature dropped, and turning it down when it rose too high.

To save us time and trouble, modern ovens are fitted with simple control devices called **thermostats** which can be set at the desired temperature and left to do the adjustment automatically. In the oven the thermostat acts as both the **sensor** and **comparator**; the **actuator** is the heating element that can bring the level of the output back to the set point (Fig F.3). The actuator needs to be linked to the sensor so that appropriate action is taken. Information about the temperature of the oven is linked to the amount of current flowing to the heating element. This information is known as *feedback*.

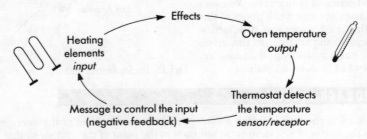

Fig F.3 Feedback; how negative feedback controls an oven

FERMENTATION

Feedback can be either *positive* or *negative*. As positive feedback leads to systems going out of control it is not useful in maintaining homeostasis.

The example above illustrates how negative feedback *can* be used to maintain a steady state. Feedback carries simple messages that are often in code. The messages being transmitted in the oven example were of two types, each *reversing* the direction of the temperature change taking place. The two messages were:

1 'The oven is too hot, reduce the current to the heating elements' and
2 'The oven is too cold, increase the flow of current'.

If the message had been: 'The oven is too hot, increase the flow of current to the heating elements', then this *positive* feedback would have resulted in very burnt food. A good example of positive feedback is the chain reaction in a nuclear reactor out of control.

Two examples of feedback in mammals are body temperature regulation and water level, both of which need to be automatically controlled (Fig F.4). ◄ Osmoregulation ►

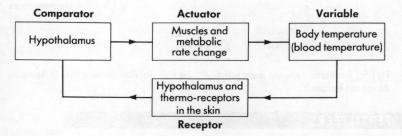

Fig F.4 Feedback; temperature control in a mammal

FERMENTATION

Micro-organisms can live and grow without the need for oxygen (anaerobic respiration). When they do this they sometimes produce products which are useful; this process is called fermentation. For example yeast, (a micro-organism) when added to fruit juices, in the absence of air digests the sugars, creating alcohol as a waste product and releasing carbon dioxide.

► FERMENTATION IN ACTION

Beer, bread and winemaking are all a result of fermentation by yeasts. Lactic acid bacteria convert milk into yoghurt and micro-organisms are needed to convert milk into cheese. Fermentation of some moulds and bacteria produce antibiotics, whilst others can produce industrial chemicals such as acetone and glycerol.

FERTILIZATION

Fertilization is the fusion or joining of the male and female **gametes** (Fig F.5a)). In mammals, the nucleus of the sperm cell penetrates the egg cell and the chromosome number is restored to its *diploid* or full number (Fig F.5b)). The fertilized egg (**zygote**) begins to develop into a new individual by **mitosis**.

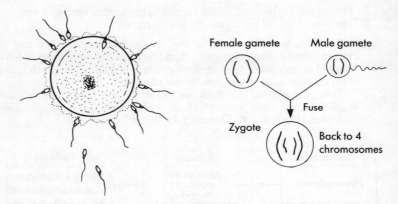

Fig F.5a) Fertilization; only one sperm enters the egg and fertilizes it

Fig F.5b) Fertilization; restoration to the diploid number

FERTILIZERS

When plants grow they take certain substances from the air (for example carbon dioxide) and from the soil (water and minerals). The supply of minerals in the soil has to be replaced each season; this can happen by adding organic manure, by growing other plants and digging them in, or by adding chemical fertilizers.

Fertilizers contain **salts** (compounds) and provide plants with the **minerals** (elements) they need. Nitrogen is by far the most important of these, followed by phosphorous and potassium.

Element	Ion	Function in plant
nitrogen	NO_3^-	nitrates in the soil provide the nitrogen which is essential for the plant to manufacture amino acids and proteins
phosphorus	PO_4^{3-}	phosphates are essential for every energy transfer within the cell
potassium	K^+	potassium makes many enzymes active
calcium	Ca^{2+}	raw material for cell walls
magnesium	Mg^{2+}	raw material for making chlorophyll
sulphur	SO_4^{2-}	raw material for making some amino acids

FILTRATION

Because these elements are needed by plants, salts containing these are manufactured in large quantities and sold as fertilizers. Examples are ammonium nitrate, ammonium phosphate and potassium chloride. Fertilizers containing these three salts are called N, P, K fertilizers because they provide nitrogen (N), phosphorus (P) and potassium (K). These are the three elements which need replacement in the soil more than others.

◄ Nitrogen cycle ►

FIBRE

◄ Balanced diet, roughage ►

FIBRE OPTICS

Anything light can travel through is called a 'medium', e.g. air, water, glass, etc. When light travels from one medium into another, it is *refracted* (bent). This is why the bottom of a swimming pool looks closer than it really is. Under certain conditions the light is refracted so much that it is *totally internally reflected*. Fibre optics make use of **refraction**.

An optical fibre consists of *two* types of glass (the *core glass fibre* and the *cladding glass fibre*) (Fig F.6). When light passes through the optical fibre it is continually being totally internally reflected, 'bouncing' along the fibre; it does not matter if the fibre is coiled or knotted, the light will still get through. The two types of glass must be very pure, ensuring that this is so is the most difficult part of the manufacturing process.

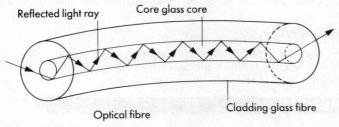

Fig F.6 Fibre optic

Optical fibres are now being used instead of copper cable to carry telephone messages. The messages are carried as pulses of light. Optical fibres are better than copper cables in that they can carry many more messages for the same thickness of cable.

FILTRATION

Filtration is a technique for separating solid particles from a liquid (or gas). This is done by passing the liquid (or gas) through a 'sieve', such as a *filter paper*. Filter paper has very tiny gaps between the fibres which make up the paper, to allow the molecules of the liquid to pass through them but *not* any solid particles which might be suspended in the liquid. For example, muddy water can be filtered using filter paper; the mud (clay particles, sand, grit etc) remaining in the paper whilst what passes through is clear water (Fig F.7). The water may still not be fit to drink, since the holes will not be small enough to trap any bacteria or harmful substances dissolved in the water.

It is a common mistake to think that substances such as salt or sugar which have been dissolved in water, making a solution, can be filtered out. This is not so; when the salt or sugar dissolves, the particles which make up these substances mix completely with the water and their particles (**ions** or **molecules**) are about the same size as the molecules of water. These are so small that they easily pass through the 'holes' in the paper.

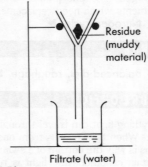

Fig F.7 Filtration

FILTRATION IN ACTION

In the home

Coffee filter papers filter the coffee solids from the coffee solution. Filters are also used in air conditioning units to remove dust particles from the air; another use is in cooker hoods, to remove grease and dust particles from the air.

In motor cars

Cars have air and oil filters, often made of paper, which remove solid particles from the air before it is drawn into the carburettor, or from the oil before it is recycled through the engine.

FLAME TESTS

A flame test is a technique which enables us to identify some of the **elements** present in substances. It is not always easy to know which **atoms** are present in a **compound**, but some atoms give distinct *colours* to flames. This is useful for making fireworks. We can also use the knowledge in the laboratory where some substances can be tested by placing a small amount on a clean wire in a 'blue' Bunsen flame. The *colour* of the flame indicates the atom present. Unfortunately, the test does not work for all atoms.

colour of flame	atom present	colour of flame	atom present
apple green	barium	lilac	potassium
orange (brick red)	calcium	yellow	sodium
green	copper	red	strontium
blue flashes	lead		

FLUORINE

FLEMING'S LEFT-HAND RULE

When an electric current is passed through a length of copper wire which is placed in the field of a strong magnet, the wire moves upwards at 90° to the direction of the magnetic field. If the current direction is reversed, then the force on the current is reversed and the wire moves in the opposite direction. The direction of force is always at right angles to the current direction *and* the field direction (Fig F.8a).

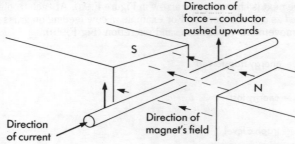

Fig F.8 a) The direction of force is at right angles to the current direction and field direction

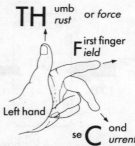

Fig F.8 b) Try this with your left hand to learn Fleming's rule for direction of current, field and force

You can learn this by using Fleming's Left Hand Rule. If the thumb and first two fingers of the left hand are held at right angles to each other, then the *thumb* gives the direction of the *force*, the *first finger* points in the same direction as the *field*, and the *second finger* points in the direction of the *current* (Fig F.8b)). An important application of this effect is the **loudspeaker**.

FLUORIDE

◀ Fluorine ▶

FLUORINE

Fluorine is a very reactive element – chemical symbol F; it belongs to a family of elements called the **halogens** (group 7 of the **periodic table**) so it has properties similar to other members of the group, for example, chlorine and bromine. Fluorine is the most reactive of the group; it is a gas and dissolves in water to form a very corrosive acid. Fluorine reacts with metals to form very stable compounds called *fluorides*. Fluoride (as sodium fluoride) is added to drinking water in some areas because fluoride has been shown to help prevent tooth decay.

FOETUS

◀ Embryo ▶

FOOD CHAIN

A food chain shows how an animal obtains its food directly from another animal or plant. The direction of transfer of energy from one **trophic level** (feeding level) to the next is shown by the arrow in Figure F.9a). At each trophic level, *energy is lost* as it is transferred. For example, a cow feeding on grass uses up energy in movement, respiration, and excretion (Fig F.9b)).

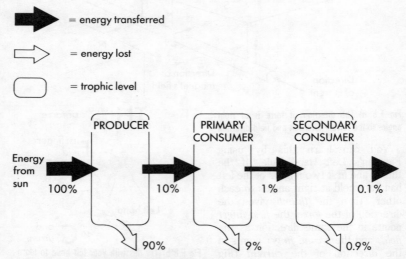

Fig F.9a) Energy is lost at each stage of the food chain

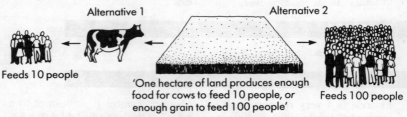

Fig F.9b) The alternative ways in which the corn produced by one hectare of land could be used

At the beginning of the food chain are the **producers** or gree plants which manufacture food by **photosynthesis**. The **consumers** obtain energy from the producers (Fig F.9c)). Feeding on the producers directly are the **herbivores** and feeding on the herbivores are the **carnivores**. **Omnivores** feed on both producers and consumers.

FOOD WEB

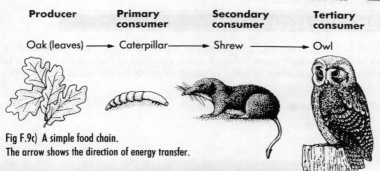

Fig F.9c) A simple food chain.
The arrow shows the direction of energy transfer.

FOOD TESTS

These tests help to identify the different type of foodstuff which may be present in a sample of food. For example, to find out if **starch** is present in potato, add *iodine solution* to the potato. The brown colour of iodine changes to blue-black, showing starch is present.

Food type	Substance used	Positive result
starch	iodine solution	blue-black colour
glucose reducing sugar	add Benedict's solution and warm tube gently	green/red colour
protein	add sodium hydroxide solution, then a few drops of copper sulphate solution	violet colour appears
fat	rub food onto filter paper	a translucent grease – stain forms

FOOD WEB

A food web is more complicated than a **food chain** as it shows how one animal may be obtaining energy from several different sources. For example, a fox may be feeding on hedgehogs, snakes, frogs, and rabbits to obtain food; a plant such as an oak tree supplies food for many different types of animals, such as snails, insects and birds (Fig F.10).

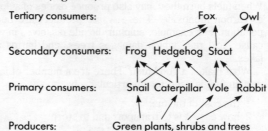

Fig F.10 Food web: in a food web one animal feeds on more than one source of food

FORCE AND ACCELERATION

When an object is at rest, a *force* such as a push or a pull must be exerted on the object to make it move. More force is required to make it go faster or to accelerate. The amount of force required depends on the mass of the object and is given the following formula:

force = mass × acceleration

f = m × a

or a = f/m

The amount of *acceleration* produced depends on two factors: the size of the force, measured in **Newtons**, and the mass of the object, measured in kilograms. If the force doubles, or the mass is halved, then the acceleration is doubled. For example, if a force of 15N acts on a mass of 3 kg then its acceleration is 5 m/s^2. If the force doubles to 30N then the acceleration is 10 m/s^2.

FOSSIL

A fossil is the remains of an organism or the shape of an organism preserved in rocks. Fossils are usually found in **sedimentary rocks** and are studied by paleontologists to find out what animals and plants may have looked like millions of years ago. The hard outer parts of the animals are usually preserved, whereas the soft tissues will have decayed before being fossilised.

FOSSIL FUELS

Coal, oil and gas are described as fossil fuels. Fossil fuels contain carbon and were formed millions of years ago by the effect of heat and pressure on decaying plants and animals. The chemical energy in the cells of the plants and animals became trapped into the coal and oil, and this energy is released as heat and light when the fuel (for example methane gas) is burned:

fuel + oxygen → carbon dioxide + water + heat

CH_4 + $2O_2$ → CO_2 + $2H_2O$ + thermal energy

When fuels burn they may also produce oxides of sulphur and nitrogen, as well as carbon monoxide. These waste products are one of the main causes of pollution. For example, sulphur dioxide dissolves in water vapour in the air to cause 'acid rain' which damages trees, and harms animal life in rivers and lakes.

What makes a good fuel? There are a number of factors which affect why a certain fuel is used for a particular job:

1 the cost of the fuel
2 how easy it is to transport and to store
3 whether it is solid, liquid or gas
4 how easily it ignites and burns

5 how much pollution is caused
6 how much energy is released when it burns

The chart below shows how much energy in millions of joules is released when 1 kilogram of fuel is burned, (the MJ means megajoules or one million joules):

Gas	55 MJ per kg
Oil	44 MJ per kg
Coal	29 MJ per kg
Wood	14 MJ per kg

Gas releases the most energy, but takes up much more space than 1 kilogram of oil, and is more bulky to transport and to store. *Oil* is easier to transport and store but costs more than gas to buy. *Wood* is much cheaper, but is bulky to store and releases less heat than the other fuels.

FRACTIONAL DISTILLATION

Fractional distillation is a process used in industry and in laboratories to separate liquids of different boiling points. Fractional distillation is used in industry to separate crude oil into 'fractions', mixtures of liquids with similar boiling points (Fig F.11). The liquids with *high* boiling points have small, light molecules. The heavier molecules need more energy to help them escape the surface of the liquid.

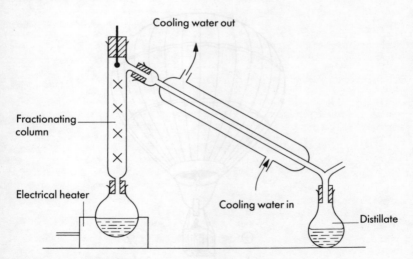

Fig F.11 Fractional distillation

One characteristic of fractional distillation is the 'fractionating column' which sits on top of the heated container. As the liquid vapourizes it travels up through the fractionating column where it continually condenses and vapourizes. This helps with the separation of liquids.

FRACTIONAL DISTILLATION IN ACTION

Separating liquids

Fractional distillation can be used to separate other liquids of differing boiling points, for example alcohol from wine.

Purifying metals

To purify zinc when it is first extracted from zinc ore. Zinc, in this form, often contains a small amount of lead. This can be separated by fractional distillation, since the boiling point of zinc is 908°C and that of lead is 1651°C.

The size and heaviness of particles is not the *only* reason for liquids to have different boiling points; the forces which hold the particles together are often more important. ◄ Evaporation ►

FREE-FALL

An object which is allowed to fall freely near the Earth's surface has a constant **acceleration** of 10 m/s^2. For example, if two objects, one of 5 kg and one of 20 kg, were dropped from a weather balloon, they would both have the same acceleration. The larger object would need more force to accelerate its greater mass (Fig F.12).

Fig F.12 Free fall; both the 5 kg object and 20 kg object would have the same acceleration

The force which is acting on free-falling objects is the force of gravity caused by the Earth's gravitational field. This is described as the weight of the object.

Since weight = mass × gravitational field strength, the weight of an object depends on how far it is from the Earth's centre. The *gravitational pull* of the Earth is decreased the further away an object is, so the weight is reduced. At the Earth's surface, the force acting on 1 kilogram is 10 Newtons. Therefore the weight of the 5 kg mass is $5 \times 10 = 50$ N, and the weight of the 20 kg mass is $20 \times 10 = 200$ N.

Since acceleration $= \dfrac{\text{force}}{\text{mass}}$

For the smaller object, $a = \dfrac{50}{5} = 10 \text{m/s}^2$

For the larger object, $a = \dfrac{200}{20} = 10 \text{ m/s}^2$ (assuming negligible air resistance)

◀ Weight ▶

FREQUENCY

The frequency of the wave is the number of complete cycles per second, measured in **Hertz** (Hz). Imagine standing on a beach and counting the waves as they come towards you. This would give you the frequency of the waves. Figure F.13 shows a wave with a low frequency and a wave with a high frequency. ◀ Speed, wavelength, frequency ▶

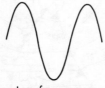
Low frequency

High frequency

Fig F.13 Frequency and wave pattern

FRICTION

When you push an object to start it moving, the object will eventually slow down and stop due to the force of *friction* which resists motion. When travelling in a car, friction between the tyres and the roads is essential if the car is going to travel safely and not skid about. Friction between the brake pads and the wheels of a bicycle is essential if you need to stop moving.

However, friction means that *energy has to be used to overcome it,* and so there are many ways of reducing friction:

1. oil is used in car engines to reduce the friction of the parts rubbing against each other;
2. a hovercraft uses air to reduce the friction between the boat and the water;
3. ball bearings are used to reduce the friction between the wheel and the axle of a skateboard.

FRONTS

There are two types of front. A *warm front* develops as warm, moist air rises up over the cold, dry air. A *cold front* develops as cold, drier air pushes underneath the warm, moist air (Fig F.14a)).

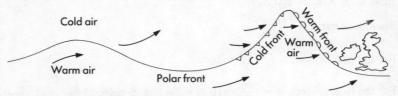

Fig F.14a) Fronts: an imaginary aerial view of the polar front

The weather map in Figure 14b) shows a typical depression or LOW, with a cold front and a warm front. As the warm front passes over, the weather pattern would show increasing wind and cloud cover, light rain or drizzle and a fall in pressure. As the cold front passes, the rain becomes much heavier, the wind becomes very strong and the temperature drops. The pressure then rises and the rain ceases as the sky clears.

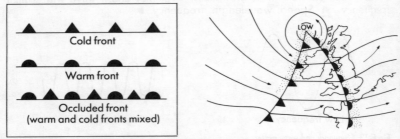

Fig F.14 b) Fronts: a typical depression or LOW will bring rain into Britain

FUNGI

◄ Micro-organism ►

FUSES

A fuse is a thin piece of wire which melts and breaks the circuit if too much current is flowing. To find out the correct fuse for a particular appliance use the following formula:

$$\text{current} = \frac{\text{watts}}{\text{volts}}$$

For example, a 60W table lamp uses $\frac{60}{240} = 0.25$A, so a 3 amp fuse would be the correct one to use. However, an electric kettle using 2000W would take a current of $\frac{2000}{240} = 8.3$ amps, so a 13 amp fuse is needed.

GALAXY

Millions of stars make up what is described as a galaxy. Our solar system is part of the galaxy known as the Milky Way.

GAMETES

Gametes are special cells in the body called sex cells. Female gametes are produced in the **ovary**, male gametes are produced in a **testis** in animals, or stamen in plants (Fig G.1). Gametes are produced by a process of cell division known as **meiosis**. Meiosis only occurs in the sex organs and reduces the chromosome number by half in each cell. All the cells produced by meiosis are different from each other. The other cells in the body are a result of cell division by *mitosis* and are identical to the parent cells with exactly the same **chromosomes**.

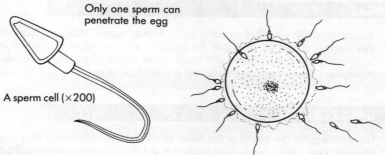

Fig G.1 Gametes

The gametes have half of the information content required so two gametes, one male and one female gamete, must fuse at fertilization to produce the first normal cell of the offspring. At fertilization each gamete carries only one of each type of chromosome. When fusion has taken place the **zygote** has two full sets of chromosomes which is normal for an ordinary body cell.
◄ Meiosis, mitosis ►

GAMMA RAYS

Gamma rays are a form of electromagnetic radiation. They are the most penetrating of the three types of radiation, **alpha**, **beta** and **gamma** and are only stopped by several centimetres of lead. They have very short wavelengths and are not deflected by electric and magnetic fields (Fig G.2).

Gamma rays are used to irradiate food in order to prolong shelf-life of food (this does **not** make the food radioactive) but can affect the taste of some foods and so is not suitable for all.

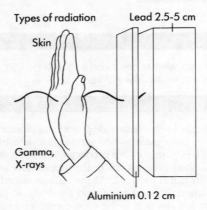

Fig G.2 Gamma radiation

GAS

All matter can be classified as either **solid**, **liquid**, or **gas**. These are called the three states of matter. In a gas the **particles** are moving very fast and are large distances apart. Gases are made up of either single **atoms** or small **molecules**. Gases exert pressure; **air** is a mixture of gases.

GAS EXCHANGE

Gas exchange in the *alveoli* of the lungs takes place by **diffusion**. Diffusion of particles takes place from where there is a higher concentration to where there is a lower concentration. Particles will diffuse until they are evenly distributed. ◄ Breathing ►

GAS LAWS

The particles in a gas are moving very fast. When these particles *hit* something they exert a *force* on that object. The combined effect of the many millions of particles in a gas on any area is its **pressure**.

Pressure is measured as force per unit area; the units are Newtons per square metre (N/m^2). The *pressure* of a gas can be increased by increasing its temperature (heating). The gas particles will have more **kinetic** energy so will be moving faster and striking the sides of a container harder and more often (Fig G.3). The pressure of a gas can also be increased by reducing the volume of that gas. The particles in the gas will then be closer together, so they will strike the walls of the container more often.

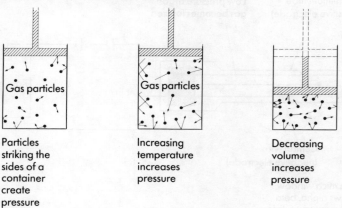

Fig G.3 Gas laws; pressure

Remember: the *volume* of a gas can be increased by increasing the temperature (expansion).

The following ideas may help you to understand the 'gas laws' or 'gas patterns':

1 pressure α temperature – provided the volume remains the same.
2 pressure α 1/volume – provided the temperature remains the same.

These two relationships can be re-expressed as:

P/T = a constant, and P×V = a constant.

In order to make calculations using these patterns or laws, temperature is measured on the **Kelvin** scale of absolute temperature. (0 Kelvin = −273° Celsius). 0K is known as *absolute zero*; it is the temperature at which particles have *no* kinetic energy.

GEIGER-MULLER TUBE

The Geiger-Muller (G-M) tube is an instrument used to detect radioactivity and measure the strength of **alpha**, **beta**, and **gamma** rays. Radiation enters the window of the G-M tube and creates *argon ions* and *electrons*. When the ions reach the electrodes, they produce a current pulse which is amplified and fed into a ratemeter from which the average pulse-rate can be determined (Fig

G.4). The G-M tube is usually connected to a **loudspeaker** which emits a series of clicks in proportion to the strength of the radiation.

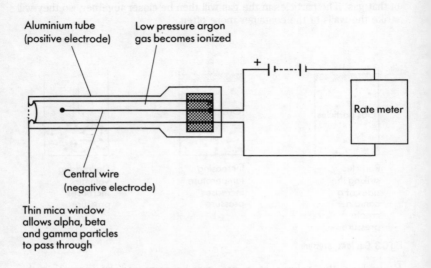

Fig G.4 Geiger-Muller tube

GENE

The instructions for a particular characteristic, or trait, are called a gene. There are genes for all our characteristics such as eye colour, hair colour and blood type. Each **chromosome** carries many genes, each a certain length of DNA. We have two copies of each gene in every normal body cell, one in each of a pair of chromosomes because we inherit one gene from our father and one from our mother of each type. These genes may be identical or have slightly different effects. For instance, we all have two genes for eye colour, if one is for blue eyes and one for brown eyes you will have brown eyes (see **dominant gene**). We can't see genes, but their effects have been observed and patterns of inheritance discovered by scientists.

It is over 100 years since an Austrian monk, Gregor Mendel, established the basic laws of inheritance. Since then the science of *genetics* has become well established. Our understanding of genetics has developed to a point where we can now create new organisms by *genetic engineering*, that is by transferring genes from one organism to another. ◀ **DNA, dominant genes, Mendel's Laws** ▶

GENERATOR

A generator transfers **kinetic energy** into electrical energy. It works on the principle that a current can be *induced* in a coil when the coil is turned in a magnetic field between the poles of two bar magnets (Fig G.5). There are basically two types of generator:

1. A DC generator, which produces one way direct current, such as bicycle dynamo.
2. An AC generator or **alternator**, which produces alternating current, such as that found in a power station and in a car.

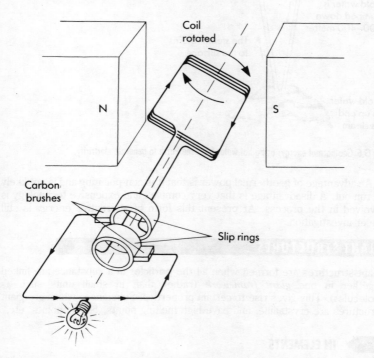

Fig G.5 Generator; the construction of a simple AC generator or alternator

GEOTHERMAL ENERGY

Geothermal energy is an example of an **alternative** source of energy. Using geothermal energy, there are basically two ways of producing the steam required to drive the turbines which generate electricity. One method, used in Cornwall, is to use the heat which is trapped in hot granite rocks deep in the Earth to heat up water and convert it into steam. Another method is to drill a deep well to release steam from hot underground water which occurs in volcanic areas (Fig G.6). In both cases the steam can then be used to drive generators and produce electricity.

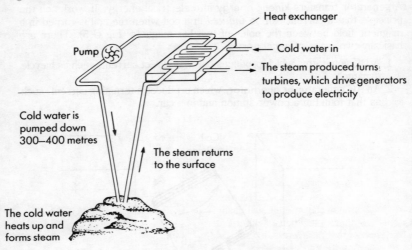

Fig G.6 Geothermal energy; using hot water from the Earth to generate electricity.

An advantage of geothermal power is that it is non-polluting and is not likely to run out. A disadvantage is that very complex and expensive technology is involved in the process. At present this form of alternative energy is still under investigation.

GIANT STRUCTURES

Giant structures are formed when all the **particles** in a substance are linked together in one *giant framework* (rather than in small units such as **molecules**). This gives rise to certain properties. Substances which are giant structures are crystalline, and have high melting points, boiling points, etc.

IN ELEMENTS

Carbon forms large numbers of covalent bonds between its atoms. It can do this in two ways to form diamond or graphite. These two forms of carbon are called **allotropes**, and both are crystalline.

Diamond is very strong because each carbon atom is linked to four others. A diamond crystal is one giant molecule. *Graphite* is very strongly bonded in layers; each layer is a giant molecule, however the forces holding the layers together are weak so they slide over each other. This property is made use of in lubricants and in pencils. The pencil 'lead' is really graphite.

Silicon is in the same group as carbon; it too has similar abilities to form giant structures. The structure of silicon is the same as that of diamond.

METALS AS GIANT STRUCTURES

Metals are giant structures. The metal atoms are 'bonded' together in an unusual way. The metal atoms lose some of their outer electrons and so become, in effect, positive ions. These electrons then move around the atoms freely. The metal atoms/ions are floating in a 'sea' of electrons. The electrons are free to move and are shared by all the atoms (Fig G.7). This idea helps to explain many of the properties of metals:

- *High boiling points and melting points*: the attraction between the 'ions' and electrons is strong, so the metals will have high melting points and boiling points and will be strong and hard.
- *Conduction of electricity*: the ease of movement of electrons will mean that metals will easily conduct electricity when a potential difference is applied across the metal.

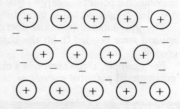

Fig G.7 Metals are giant structures

COMPOUNDS

Any **compound** containing **ions** will form giant structures; each positive ion surrounds itself with as many negative ions as possible and vice versa. The way in which the ions pack together depends on the size of the individual ions and their charge, i.e. 1+, 2+ or 1−, 2−, etc.

In addition, *silicon dioxide*, a compound of silicon, has a giant structure, the atoms being bonded by **covalent bonds**. We come across this substance quite often; it appears as sand, as quartz in rocks and can be made into glass. It is in fact the second most common element found in the Earth's crust (28%), the first being oxygen.

GLASS

Glass is a very useful material, it is resistant to corrosion, is waterproof, does not conduct electricity, and can be made transparent. The raw materials are cheap and easily available, such as silica (sand). Glass is made by heating a mixture of oxides and carbonates in a furnace, the exact combination depending on the type of glass required. The molten material is then moulded and cooled.

GLUCOSE

Glucose is a type of **sugar**; its chemical formula is $C_6H_{12}O_6$. Glucose is the sugar that plants first produce in the process of **photosynthesis**, and is then converted to more complex molecules, like sucrose (another sugar: the sugar we buy from a shop) and starch − a **polymer** of glucose molecules.

GRANITE

A hard, **igneous** rock containing different minerals, such as quartz, felspar and mica.

GRAVITATIONAL POTENTIAL ENERGY

When an object is at rest at a point above the ground it has gravitational potential energy. If you hold a book above your desk, the book has gravitational potential energy equal to the amount of work which you did to lift it to that height, which is given by force × distance moved.

The gravitational potential energy of an object is mgh, where mg is the *upward force* needed to lift the object, and h is the *vertical height* above the ground.

For example, if the book has a mass of 2 kg, and is held 3 m high, then the gravitational potential energy = $2 \text{kg} \times 10 \text{ m/s}^2 \times 3 \text{ m} = 60 \text{ J}$.

GRAVITY

Gravity is the attraction of the Earth for solids, liquids and gases. This means that if you drop an object it will always fall, due to the attraction of the Earth for the object. The Moon has one sixth the gravitational attraction of the Earth as it is smaller in size.

The apparatus in Figure G.8 can be used for *measuring* the acceleration due to gravity. ◀ Weight ▶

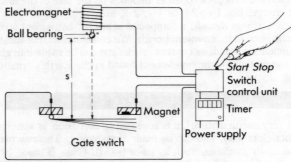

Fig G.8 Apparatus for measuring the acceleration due to gravity

GREENHOUSE EFFECT

▶ CAUSES

The Earth is surrounded by an **atmosphere** which acts as a 'blanket' keeping it warm. On Earth the average surface temperature is +15°C, whereas on the Moon, which has no atmosphere, the surface temperature is −18°C.

GREENHOUSE EFFECT

Evidence has shown that the Earth is slowly warming up, caused by changes in the atmosphere. This gradual warming is called the 'Greenhouse effect'.

During the last century man has been adding to the amount of *carbon dioxide* in the atmosphere as a result of buring fossil fuels (coal, oil, gas) and burning trees (as a result of deforestation to provide land for growing crops). This has disturbed the balance of carbon dioxide in the atmosphere; it has been estimated to have increased over the last century from a 'pre-industrial' level of 0.275% to 0.350% today. This is turn has resulted in a slight warming of the Earth. In addition, other man-made gases have been released into the atmosphere which have been shown to add to the problem, for example **chlorofluorocarbons** (CFCs), methane, nitrous oxide and ozone. (Be careful – this is an increase in ozone in the lower atmosphere – do not confuse it with the **ozone layer**.)

How the greenhouse effect works

The Earth is warmed by solar radiation which passes through the atmosphere and is absorbed by the ground and oceans, warming them up. This energy is re-radiated (at a different wavelength) as *heat* into the atmosphere. Carbon dioxide, water vapour and the other gases mentioned absorb this heat energy and then re-radiate it *back* to the surface. Any increase in the amount of carbon dioxide will therefore increase the amount of heat 'trapped', resulting in a gradual warming (Fig G.9).

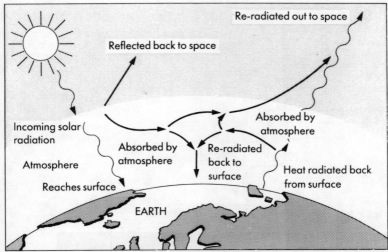

Some of the energy radiated at infrared wavelengths from the ground is absorbed and re-radiated downwards by the atmosphere – the greenhouse effect

Fig G.9 The greenhouse effect

GUT

IMPLICATIONS

Many predictions have been made as to the effect that a slight warming of the Earth (1°C to 2°C) will produce, using many different computer models. Some of the suggestions put forward include:

- a change in climate; some areas will become dryer, some wetter, some warmer. Droughts may occur in parts of the Earth and floods in others.
- Polar ice caps will melt, and this will itself affect the climate.

No one is certain what will happen – only that there will be changes.

GUT

◀ Intestine ▶

HABITAT

The habitat is the place where an **organism** lives. For example, the habitat of earthworms is in the soil, the habitat for a crab is on a rocky beach.

HAEMATITE

Haematite is an ore of iron that is red in colour and is sometimes referred to as kidney ore because some samples are shaped like kidneys. It is an oxide of iron, chemical composition Fe_2O_3. The iron is extracted from the ore by **reduction** in the **blast furnace**.

HALF-LIFE

After a period of time any radioactive material will decay and become stable (non-radioactive). You cannot predict, however, when any particular nucleus will decay and release its radiation and energy. They decay in a *random* way. Different substances do however decay at different rates. The decay rate of any radioactive material is measured in Half-lives.

The *half-life* of a radioactive material is the time taken for the radioactivity to reduce by half. For any particular radioactive material, the half-life is constant whatever the conditions, since radioactive decay is unaffected by temperature or pressure. Half-lives can be very long or very short (thousands of years to less than a second). For example, the half-life of carbon-14 is 5570 years, but for carbon-10 it is only 19 seconds. If you plot a graph for the decay of any radioactive material it will always follow the same pattern (Fig H.1).

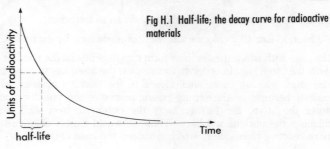

Fig H.1 Half-life; the decay curve for radioactive materials

HALOGENS

The halogens are a family of **elements** (group 7 of the **periodic table**) which have similar chemical properties (Fig H.2).

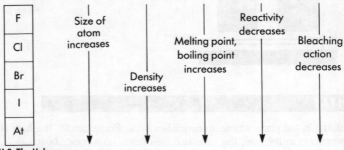

Fig H.2 The Halogens

TYPES OF REACTION

Reaction with metals

The halogens will react with metals to form metal halides, for example iron will react with chlorine:

$$2Fe + 3Cl_2 \rightarrow 2FeCl_3$$

The reactivity of the halogens decreases as you move down the group.

Reaction with water

The halogens will react with water to form acidic solutions which also act as bleaches. In the same way that the reactivity *decreases* as you *move down* the group, so does the bleaching power of the solutions. The solutions produced from iodine and bromine are weak acids, whereas chlorine and fluorine will produce strong acids:

$$H_2O + Cl_2 \rightarrow HCl + HOCl \quad \text{(chloric acid – bleach)}$$

REACTIVITY TRENDS

When halogens react with metals they do so to form ions:

F^-, fluoride ion; Cl^-, chloride ion; Br^-, bromide ion; I^-, iodide ion.

The ease with which these atoms form ions depends on the number of *electron shells* the atom has. In order to form an ion the atom has to *gain* an electron. The atom with its outer shell closer to the positive nucleus will find this easiest, because of the strong pulling-power of the positive nucleus. The larger the atom, the further away the outer electron shell, so the less influence the nucleus will have. Thus we would expect chlorine to be much more reactive than iodine, which is indeed the case (Fig H.3).

HARD WATER

	Chlorine	*Bromine*	*Iodine*
At room temperature	gas	liquid	solid
Reaction with iron	very fast	fast	slow
Reaction with potassium iodide solution	reacts	reacts	no reaction
Effect on indicator paper	bleaches	bleaches	bleaches

Fig H.3 Halogen reactivity

HARD WATER

Tap water often contains dissolved salts, depending on what type of rock the rain water has washed through. Different parts of the country will have different salts and different amounts of salts dissolved in their water.

Hard water is water that has either magnesium or calcium salts dissolved in it (it contains Mg^{2+} or Ca^{2+} ions). The problem with hard water is that the calcium ions or magnesium ions it contains will react with soaps. These ions attach themselves to soap molecules, forming an insoluble substance which appears as scum on the surface of the water. If the soap is no longer in the water then it cannot do its job of emulsifying grease. More soap has to be added to hard water than to soft water in order to do the same job.

Soapless detergents do **not** behave in this way so there is then no scum formed with hard water and hence no wasted detergent.

REMOVING HARDNESS IN WATER

Hardness in water can cause problems which can be more serious than just wasting soap, so it is important to be able to remove this hardness.

The salts which cause hardness are calcium hydrogen carbonate, $Ca(HCO_3)_2$ (formed from limestone rock, water and carbon dioxide); calcium sulphate, $CaSO_4$ (from gypsum rock); magnesium hydrogen carbonate, $Mg(HCO_3)_2$; and magnesium sulphate, $MgSO_4$.

When water containing hydrogen carbonates of calcium or magnesium are heated, a chemical reaction occurs producing insoluble carbonates. This is the 'fur' (limescale) inside hot water pipes and kettles:

$$Ca(HCO_3)_2 \rightarrow CaCO_3 + H_2O + CO_2$$

This 'furring-up- of pipes can cause damage and reduces the efficiency of hot-water-heating systems.

Hardness due to these salts can be removed by boiling but this is not always convenient. Other methods of removing hardness, include:

1 Adding water softeners; these are chemicals which can be added to remove calcium ions from solution
2 Using ion-exchange resins; these are in water softeners which are built into many washing machines. They work by replacing calcium ions with sodium ions, so they need to be regularly topped up with salt (sodium chloride).

HEART

The heart is basically two muscular pumps which work side by side. Each side is divided into two chambers, an *upper atrium* and a *lower ventricle*. (Fig H.4). The right atrium takes in deoxygenated blood which has been round the body, and the right ventricle pumps the blood to the lungs, via the **pulmonary artery**. The left atrium takes in oxygenated blood from the lungs, via the **pulmonary vein**, and the more muscular left ventricle pumps the blood under great pressure around the body, via the aorta, the thick walled main artery (Fig H.5).

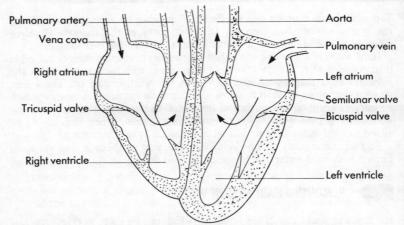

Fig H.4 The heart is a powerful pump which pumps blood round the body and to the lungs

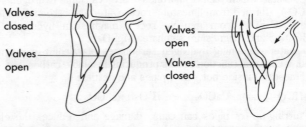

a) Ventricle relaxed: blood is forced from the atrium into the ventricle. Valves prevent blood flowing 'backwards'

b) Ventricle contracted: blood is forced from the ventricle and out of the heart. The atrium meanwhile re-fills with more blood

Fig H.5 The heart in action

It is important to learn the sequence of blood flow through the heart (Fig H.6).

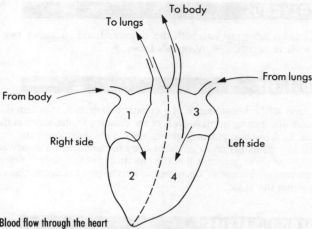

Fig H.6 Blood flow through the heart

HELIUM

Helium (chemical symbol He) is a gaseous element found in the air. It is less dense than air so is often used to fill balloons. It belongs to a family of elements called the **inert gases** (group 0 of the **periodic table**) and is very non-reactive. It is also used by underwater divers as a replacement for nitrogen in their air tanks. The problem with nitrogen is that as the diver begins to surface the reduction in pressure can cause bubbles of nitrogen to develop in the blood, causing 'the bends' a very painful and dangerous complaint. Helium does not do this, but it does however cause one added problem in that divers breathing helium-rich air sound a little like 'Donald Duck' when they speak.

A *helium nucleus* contains two protons and two neutrons. Helium nuclei are emitted by some radioactive materials, they are known as **alpha particles**.

HERBIVORE

A herbivore is an animal which obtains its energy by feeding directly on green plants. For example, sheep, cows and rabbits are herbivores. ◀ Food chain, food web ▶

HERMAPHRODITE

◀ Sexual reproduction ▶

HERTZ

The hertz is the S.I. unit for frequency. One hertz is one cycle every second. The mains current supplied to houses in Europe has a frequency of 50 hertz (50 Hz).

HETEROZYGOUS

This means an individual has both the **dominant** and **recessive gene** for a particular characteristic. ◄ Mendel's Laws ►

HOMEOSTASIS

Homeostasis is the maintenance of a constant environment within the body. This means maintaining normal levels of the dissolved substances in the blood and tissue fluid, such as glucose, amino acids, salts, hormones and excretory products, as well as pH and a constant body temperature. The advantage of homeostasis to the organism is that the functioning of the body is independent of the external environment. **Osmoregulation** helps to maintain this constant balance within the blood.

HOMOLOGOUS SERIES

A homologous series describes a group of organic compounds which have similar chemical properties: for example they can be represented by a general formula; they can be made by similar reactions; and there is a *regular change* in their physical properties – e.g. their melting points increase as the relative molecular matter increases.

The name *ending* shows the homologous series of a compound, as can be seen in the chart below:

Functional group	Name ending	Homologous series
C–C	-ane	alkane
C=C	-ene	alkene
C–O–H	-ol	alcohol
CO_2H	acid	acids

The number of **atoms** is indicated by the first part of the name, as can be seen in the next chart:

Number	Start of name
1	meth-
2	eth-
3	prop-
4	but-
5	pent-

Examples of the members of each homologous series are shown in Figure H.7a)–c). ◄ **Hydrocarbon** ►

Alkanes	(General formula: C_nH_{2n+2})
methane	CH_4
ethane	C_2H_6
propane	C_3H_8
butane	C_4H_{10}

a) Alkanes

Alkenes	(General formula: C_nH_{2n})
ethene	C_2H_4
propene	C_3H_6
butene	C_4H_8
pentene	C_5H_{10}

b) Alkenes

Alcohols	(General formula $C_nH_{2n+1}OH$)
methanol	CH_3OH
ethanol	C_2H_5OH
propanol	C_3H_7OH
butanol	C_4H_9OH

c) Alcohols

Fig H.7 Homologous series

HOMOZYGOUS

This means an individual has two *identical* **genes** for a particular characteristic, either both **dominant** genes or both **recessive** genes.
◄ Mendel's Laws ►

HORMONE

Hormones are chemical substances produced in very small amounts by special glands in the body called *endocrine glands*. Hormones are carried from the gland via the blood stream to their target organ where they have their effect. For example, **oestrogen** is produced in the **ovaries** but has its effect on the development of the breasts and growth of pubic hair. The *thyroid gland* produces a hormone called *thyroxine* which controls the rate of metabolism in humans, and therefore controls rate of growth. The thyroid gland is controlled by the 'master gland' of the body, the **pituitary**.

Plants also produce hormones in very small amounts and these help to control the growth of the plant.

HYBRIDS

If two different varieties of animals or plants, each of which has useful characteristics, are allowed to breed together, the offspring are known as *hybrids* and will possess the characteristics of both the varieties that were cross-bred. The technique of cross-breeding has been used to great advantage in producing disease-resistant plants and high-yielding crop plants. Hybrids such as varieties of corn were introduced to America in the 1930s and resulted in increased yields of up to 50%. You will find that seeds of hybrid flowers and vegetables are very expensive due to the amound of research that has been done in developing the hybrid seeds. ◄ Mendel's Laws ►

HYDROCARBON

Hydrocarbons are compounds which contain carbon and hydrogen atoms only. There are different groups of hydrocarbons which form a **homologous series**.

Alkanes

The **alkanes** have the general formula C_nH_{2n+2}. Some typical alkanes, with their formula, are methane, CH_4; ethane, C_2H_6; propane, C_3H_8; and butane, C_4H_{10}.

Alkenes

The **alkenes** have the general formula C_nH_{2n} and contain one carbon – carbon double bond (C=C) in their molecule. Some typical alkenes, with their formula, are ethene, C_2H_4; propene, C_3H_6; butene, C_4H_8; and pentene, C_5H_{10}.

Alkynes

The alkynes have the general formula of C_nH_{2n-2} and contain one carbon-carbon triple bond (C≡C) in their molecules. Some typical alkynes, with their formula, are ethyne, C_2H_2; propyne, C_3H_4; butyne, C_4H_6; and pentyne, C_5H_8.

As one may suspect, the main source for these compounds is *crude oil* which is a mixture of hydrocarbons. Hydrocarbons are used as fuels and as raw materials for the manufacture of **plastics**. Petrol is a mixture of hydrocarbons, the proportion of each determining the petrol's *octane rating*.

HYDROCHLORIC ACID

Hydrochloric acid is a strong acid, chemical formula HC1. It reacts as a typical acid, (for example, in dilute form it reacts with metals and neutralizes bases, etc.). It is used in industry for cleaning metals, and in the manufacture of printed circuits and dyes.

HYDRO-ELECTRIC POWER (HEP)

The movement of fast flowing rivers through turbines in a hydro-electric power station produces electricity. The **kinetic energy** of the fast-flowing river is used to turn a water turbine. The turbine is connected to a dynamo which generates electricity. Some hydro-electric power stations work on a system of two reservoirs, one higher than the other. The water is pumped back up to the high reservoir from the lower reservoir during the night when electricity is cheaper (Fig H.8).

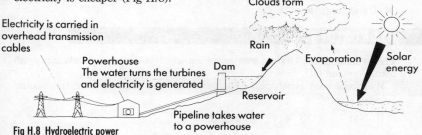

Fig H.8 Hydroelectric power

Two advantages of HEP is that there is no pollution, and the source of energy is unlikely to run out. One disadvantage is that the power station may be very costly to build and maintain. Sometimes the power stations may spoil the appearance of the environment.

HYDROGEN

Hydrogen is a gaseous element, chemical symbol H. It is the simplest atom possible, containing 1 **proton** in its **nucleus**. Hydrogen is a gas which is lighter than air, so it was used in the first airships. The only problem is that it is extremely inflammable, and this property led to the *Hindenburg* disaster in which a great airship of the 1930s caught fire.

Hydrogen gas is given off in any reaction between metals and dilute acids. It is a colourless, odourless gas but can be recognized by the characteristic 'pop' it gives when a lighted splint is held at the mouth of a test tube containing hydrogen. The hydrogen of course is exploding. Hydrogen is also used as a fuel in rockets.

HYDROLYSIS

Hydrolysis is a type of chemical reaction in which a **molecule** of water reacts with a **reactant**. One important hydrolysis reaction is that of starch (a polymer), which is broken down into *sugars* (small molecules). The hydrolysis product of starch with an acid catalyst is *glucose*:

starch + water → glucose

If the hydrolysis of starch is catalysed by the enzyme **salivary amylase**; the product is *maltose* (another sugar) This difference in product highlights the importance of the catalyst in chemical reactions:

starch + water → maltose

◀ Digestion, proteins ▶

HYDROPHILIC

◀ Detergents ▶

HYDROPHOBIC

◀ Detergents ▶

HYDROXIDE ION

The hydroxide ion, OH$^-$, is a negative ion which is the cause of alkalinity. All **alkalis** contain OH$^-$ ions.

Hydroxide ions react with hydrogen ions from acids to form the neutral substance *water* :

H$^+$(aq) + OH$^-$(aq) → H$_2$O(l)

Most hydroxides are insoluble, but those of sodium, potassium, lithium and ammonium are soluble.

HYPOTHALAMUS

The hypothalamus is part of the mammalian brain which responds to temperature changes both inside and outside the body. The temperature of the blood flowing through the brain is monitored by the hypothalamus and information about the external temperature is detected by special *thermoreceptors* in the skin, which are connected by nerves to the hypothalmus. The brain initiates responses which are appropriate to the information received. If the temperature is too high:
1 the body is cooled by sweating;
2 the hair lies flat against the skin;
3 blood is pumped to capillaries just below the skin surface;
4 there is a general lowering of the body's metabolic rate.

A fall in temperature would cause the opposite responses plus shivering to raise the temperature by producing heat in the muscles. Mammals are very sensitive to temperature change and humans soon die if the body core temperature is too high or too low. ◀ **Vasoconstriction, vasodilation, body temperature regulation** ▶

IGNEOUS ROCKS

Igneous rocks are formed when very hot molten magma from the mantle cools and solidifies. The size of the crystals which can be seen in igneous rocks indicates the rate of cooling of the magma. Small crystals are formed when the magma cools rapidly. Granite and basalt are examples of igneous rocks.

INDUCED CURRENT

When a coil of wire is moved in a magnetic field or there is a change in the magnetic field around a coil then a current is *induced* in the coil. Faraday explained this effect by suggesting that an **e.m.f.** was induced in a conductor whenever it cuts magnetic field lines (Fig I.1). Three factors *increase* the induced e.m.f:

1 how fast the magnet or coil is moved;
2 how many turns there are on the coil;
3 how strong the magnet is.

The size of the induced e.m.f. is directly proportional to the rate at which the conductor cuts the magnetic field lines.

◄ Electromagnetism ►

Fig I.1 Induced current

G = galvanometer

INFRA-RED RADIATION

These are **electromagnetic waves** which are radiated by all warm objects including the Sun. The infra-red rays are invisible and their wavelength is just longer than that of visible red light. The upper levels of the **atmosphere** absorb some of the infra-red radiation from the Sun to prevent the surface of the **Earth** from becoming too warm.

INHERITANCE

Offspring inherit characteristics from both parents as each pair of chromosomes in the nucleus of the cell contains one **chromosome** from the male parent and one from the female parent. Carried on the chromosomes are the **genes** which determine the appearance of the offspring. ◄ Mendel's Laws, gametes, sexual reproduction, variation ►

INORGANIC COMPOUNDS

The compounds are listed in alphabetical order.
The following abbreviations are used:

- dec means that the compound decomposes before it melts or boils
- h means that the crystals are often hydrated so that they give off water on heating before they melt
- sub means that the compound sublimes

The solubility is indicated approximately using these abbreviations:

- i insoluble
- sl.s slightly soluble
- s soluble
- vs very soluble
- r means that the compound reacts with water

All the compounds are white or colourless unless otherwise stated. Some properties of inorganic compounds are shown in the chart below.

◀ Compounds, organic compounds ▶

Compound	Formula	Structure	Melting-point °C	Boiling-point °C	Solubility	Notes
aluminium chloride	$AlCl_3$	giant (ions)	sub		r	
aluminium hydroxide	$Al(OH)_3$	giant (ions)	300	dec	i	
aluminium oxide	Al_2O_3	giant (ions)	2015	2980	i	
ammonia	NH_3	molecular	-78	-34	vs	
ammonium chloride	NH_4Cl	giant (ions)	sub		s	
ammonium nitrate	NH_4NO_3	giant (ions)	170	dec	vs	
ammonium sulphate	$(NH_4)_2SO_4$	giant (ions)	dec		s	
barium chloride	$BaCl_2$	giant (ions)	963	1560	s	h
barium sulphate	$BaSO_4$	giant (ions)	1580		i	
calcium carbonate	$CaCO_3$	giant (ions)	dec		i	
calcium chloride	$CaCl_2$	giant (ions)	782	2000	s	h
calcium hydroxide	$Ca(OH)_2$	giant (ions)	dec		sl.s	
calcium nitrate	$Ca(NO_3)_2$	giant (ions)	561	dec	vs	h
calcium oxide	CaO	giant (ions)	2600	3000	r	
carbon monoxide	CO	molecular	-205	-191	i	h, green
carbon dioxide	CO_2	molecular	sub		sl.s	h, blue
copper(II) chloride	$CuCl_2$	giant (ions)	620	dec	s	
copper(II) nitrate	$Cu(NO_3)_2$	giant (ions)	114	dec	vs	
copper(I) oxide	Cu_2O	giant (ions)	1235		i	red
copper(II) oxide	CuO	giant (ions)	1326		i	black
copper(II) sulphate	$CuSO_4$	giant (ions)	dec		s	h, blue
hydrogen bromide	HBr	molecular	-87	-67	vs,r	
hydrogen chloride	HCl	molecular	-114	-85	vs,r	forms hydrochloric acid in water
hydrogen iodide	HI	molecular	-51	-35	vs,r	
hydrogen oxide (see water)						
hydrogen peroxide	H_2O_2	molecular	0	150	vs	
hydrogen sulphide	H_2S	molecular	-85	-60	sl.s	

INORGANIC COMPOUNDS

Compound	Formula	Structure	Melting-point °C	Boiling-point °C	Solubility	Notes
iron(II) chloride	$FeCl_2$	giant (ions)	667	sub	s	yellow–green
iron(III) chloride	$FeCl_3$	giant (ions)	307	dec	s	h, orange
iron(III) oxide	Fe_2O_3	giant (ions)	1565		i	red
iron(II) sulphate	$FeSO_4$	giant (ions)	dec		s	pale green
lead(II) chloride	$PbCl_2$	giant (ions)	501	950	sl. s	
lead(II) nitrate	$Pb(NO_3)_2$	giant (ions)	dec		s	
lead(II) oxide	PbO	giant (ions)	886	1472	i	yellow
lead(IV) oxide	PbO_2	giant (ions)	dec		i	brown
lead(II) sulphate	$PbSO_4$	giant (ions)	1170		i	
magnesium carbonate	$MgCO_3$	giant (ions)	dec		i	
magnesium chloride	$MgCl_2$	giant (ions)	714	1418	s	h
magnesium nitrate	$Mg(NO_3)_2$	giant (ions)			vs	h
manganese(IV) oxide	MnO_2	giant (ions)	dec		i	black
manganese(II) sulphate	$MnSO_4$	giant (ions)	700	dec	s	h, pink
concentrated nitric acid	HNO_3	molecular	−42	83	vs, r	
nitrogen hydride (see ammonia)						
nitrogen oxide	NO	molecular	−163	−151	sl. s	
nitrogen dioxide	NO_2	molecular	−11	21	s	brown
potassium bromide	KBr	giant (ions)	730	1435	s	
potassium chloride	KCl	giant (ions)	776	1500	s	
potassium hydroxide	KOH	giant (ions)	360	1322	vs	
potassium iodide	KI	giant (ions)	686	1330	vs	
potassium manganate(VII)	$KMnO_4$	giant (ions)	dec		s	purple
potassium nitrate	KNO_3	giant (ions)	334	dec	vs	
silicon dioxide	SiO_2	giant (atoms)	1610	2230	i	
silver bromide	$AgBr$	giant (ions)	432	dec	i	pale yellow
silver chloride	$AgCl$	giant (ions)	455	1550	i	
silver iodide	AgI	giant (ions)	558	1506	i	yellow
silver nitrate	$AgNO_3$		212	dec	vs	
sodium bromide	$NaBr$	giant (ions)	755	1390	s	
sodium carbonate	Na_2CO_3	giant (ions)	851	dec	s	h
sodium chloride	$NaCl$	giant (ions)	808	1465	s	
sodium hydrogencarbonate	$NaHCO_3$	giant (ions)	dec		s	
sodium hydroxide	$NaOH$	giant (ions)	318	1390	s	
sodium nitrate	$NaNO_3$	giant (ions)	307	dec	vs	
sodium sulphate	$NaSO_4$	giant (ions)	890		s	
sulphur dioxide	SO_2	molecular	−75	−10	vs, r	
sulphur trioxide	SO_3	molecular	17	43	r	
concentrated sulphuric acid	H_2SO_4	molecular	10	330	vs, r	
titanium(IV) oxide	TiO_2	giant (ions)	1830		i	
zinc oxide	ZnO	giant (ions)	1975		i	yellow when hot
zinc sulphate	$ZnSO_4$	giant (ions)	740	dec	s	
water	H_2O	molecular	0	100		

INSULATION

During cold weather, about one third of the energy produced in Britain is used to heat people's homes to a comfortable temperature of about 20°C. But a percentage of this energy is *lost* through windows, walls, doors and the roof. This loss of heat is reduced by using various ways of trapping air to *insulate* the house. Air is a poor conductor of heat and prevents heat escaping from the house. There are four main ways of insulating a house to reduce this loss of heat:

1. Roof insulation – by putting a layer of insulating fibre material in the loft of the house. The insulation material traps air in tiny spaces between the fibres.
2. Double glazing of windows – putting a second pane of glass in each window so that a layer of air is trapped inside the two panes, which greatly reduces the amount of heat which can escape. However, most newly built houses have windows which are installed as sealed double-glazed units. These consist of two panes of glass which prevent heat escaping (Fig I.2a)). You may have had 'double glazing' put into some existing windows at home. The old single pane windows are replaced by double pane window units.
3. Cavity wall insulation – many recently built houses have walls made of two layers of bricks with a gap in between them. The gap can be filled with insulating foam which traps pockets of air and prevents heat escaping from the house (Fig I.2 b)).
4. Draught excluders – strips of draught-excluding material can be put around doors and windows to prevent warm air escaping and stop cold air from coming into the room. However, if you burn a fuel such as coal, wood or gas you should make sure that there is a good flow of air to prevent the build up of poisonous fumes.

Fig I.2 Insulation

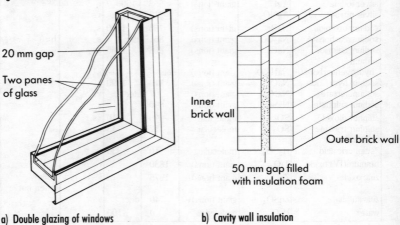

a) Double glazing of windows b) Cavity wall insulation

Some information on how heat losses can be reduced in the home, and how much money in heating bills can be saved, is shown in Figure I.2c).

Method of reducing heat loss	% of heat saved	Typical cost in £'s	Approx. time to recover cost (years)
Double glazing	15	1000	30
Carpet underlay	20	200	8
Draught proofing	25	50	1
Roof insulation	30	150	3
Cavity wall insulation	35	300	5

Fig I.2c) Reducing heat loss in the home

INTERCOSTAL MUSCLES

◀ Breathing ▶

INTESTINE

The intestine plays a vital part in *egestion* or *defecation*, where undigested material is passed out of the body through the *anus*. The large intestine is often referred to as the *colon*. Faeces are temporarily stored in the *rectum* and then egested at intervals.

IODINE

Iodine is a solid element which exists as black purplish crystals. The chemical symbol for iodine is I. Iodine is unusual because when it is heated it changes to a gas, not to a liquid. This is called **sublimation**. Iodine belongs to a family of elements called the **Halogens** (group 7 of the **periodic table**). It can be dissolved in potassium iodide solution to form 'tincture of iodine' a purple-coloured liquid which is used as an antiseptic.

ION

This is a **particle** that carries an electrical charge, which might be positive or negative. Each ion has a name and formula. Ions can be derived from single atoms or from combinations. The charge on the ion is shown as a + or − at the top of the symbols. The size of the charge is indicated as a number, e.g. 1+, 2+, 3+ etc., or 1−, 2−, 3− etc. Examples of ions and their formulae are oxide, O^{2-}; chloride, Cl^-; copper, Cu^{2+}; carbonate, CO_3^{2-}.

In **compounds** which contain ions, the *overall charge* is zero; the positive charge balances the negative charge. This helps when working out the formula for a compound; for example a compound made from *sodium ions* and *chloride ions* has the formula NaCl. The formula for *sodium carbonate* is Na_2CO_3; because the carbonate ion has a charge of 2−, two sodium ions Na+ are needed to *balance* this charge.

IONIC BONDING

IONS AND THE PERIODIC TABLE

- **Group 1** elements form ions with **one** positive charge (they have one electron to lose).
- **Group 2** elements form ions with **two** positive charges (they have two electrons to lose).
- **Group 3** elements form ions with **three** positive charges (they have three electrons to lose).
- **Group 7** elements form ions with **one** negative charge (they have one space to fill).
- **Group 6** elements form ions with **two** negative charges (they have two spaces to fill).

Some examples of ions formed from metals and non-metals are shown in Figure I.3.

Metal atoms	Group	Electrons lost	Ion formed
lithium	1	1	Li^+
sodium	1	1	Na^+
potassium	1	1	K^+
magnesium	2	2	Mg^{2+}
calcium	2	2	Ca^{2+}
aluminium	3	3	Al^{3+}

Non-metal atoms	Group	Electrons gained	Ion formed
oxygen	6	2	O^{2-}
sulphur	6	2	S^{2-}
chlorine	7	1	Cl^-
bromine	7	1	Br^-
iodine	7	1	I^-

Fig I.3 Some atoms and their ions

IONIC BONDING

There is a stable arrangement for electrons in atoms. This occurs when an atom has a *filled outer shell*. All the atoms in group 0 (**inert gases**) of the **periodic table** have filled outer shells. These atoms do *not* react with other substances, except for a few special cases. All other atoms react in order to fill their outer electron shells. They can do this in two ways: by *sharing electrons* forming **covalent bonds**; or by *transferring electrons* forming **ions**.

ELECTRON TRANSER TO FORM IONS

In the reaction between sodium and chlorine atoms both the ions that are formed have filled outer shells (Fig I.4):

Na + Cl → Na^+ + Cl^-

IONIC BONDING

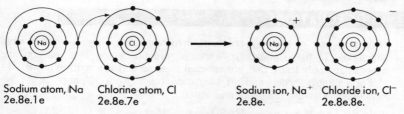

Sodium atom, Na Chlorine atom, Cl   Sodium ion, Na$^+$ Chloride ion, Cl$^-$
2e.8e.1e   2e.8e.7e   2e.8e.   2e.8e.8e.

Fig I.4 Ionic bonding: formation of ions when sodium reacts with chlorine

The sodium ion has *lost* an electron – we show this as Na$^+$; the chloride ion has *gained* an electron – we show this as Cl$^-$. Once these two ions have been formed they will *attract* each other because of their opposite charges. (Like charges repel; unlike charges attract.)

The reason the electron transfer takes place in this direction is because any transfer of electrons takes energy. It is easier to take 1 electron from sodium than to take 7 from chlorine. This results in the general rule: Metals form *positive* ions; non-metals form *negative* ions.

Note: positive ions are called **cations**; negative ions are called **anions**.

PROPERTIES OF IONIC COMPOUNDS

- Sodium ions
- Chloride ions

Fig I.5 Ionic bonding; the sodium chloride lattice

Because of their strong attraction for each other, these ions form a giant *ionic lattice*, in which each ion is surrounded by as many ions of the opposite charge as possible (Fig I.5).

The strong forces of attraction between these ions are referred to as ionic bonds. Such ionic substances have high melting points and high boiling points and are solids at room temperature (Fig I.6). Ionic substances will also usually dissolve in water.

Property	Ionic compounds	Covalent compounds
relation to periodic table	formed between metal atoms and non-metal atoms	formed between non-metal atoms
melting point	high>250°C	low<250°C
boiling point	high>500°C	low<500°C
electrical conductivity	good conductor when molten or in solution	non-conductors
solubility in water	usually soluble	usually insoluble

Fig I.6 Comparing ionic compounds with covalent compounds

Since ionic compounds contain charged particles they will conduct electricity, but only if the ions are free to move. This can happen if a) the compound is heated until it is molten, or b) the compound is dissolved in water.

IONIC EQUATIONS

This is a way of showing reactions involving ions. They are called ionic equations. In these equations, ions that are unaffected in a reaction are ignored and only those that are affected in some way are written down.

For example, when sodium hydroxide reacts with sulphuric acid, the products are sodium sulphate and water:

$$2NaOH + H_2SO_4 \rightarrow Na_2SO_4 + 2H_2O$$

In this example the sulphate ion (SO_4^{2-}) and the sodium ion (Na^+) are *unaffected* so we can *ignore* them. We can rewrite the example as:

$$OH^-(aq) + H^+(aq) \rightarrow H_2O(l)$$

This is the ionic equation that represents the reaction. It is also the general pattern for all *neutralization* reactions.

INDICATORS

An indicator is a substance, often derived from plant dyes, which tells us whether something is **acid** or **alkaline** by changing colour.

The most commonly used indicator is **Universal indicator**, since it not only tells us if something is acid or alkaline, but also if the acid (or alkali) is strong or weak. It can be used as a liquid (usually green) or soaked onto a type of blotting paper and used as a paper. The paper, of course, has to be wet in order to work since acids only behave as acids in solution.

The strength of an acid or alkali is measured on the **pH scale**. The pH scale has a range of 1 to 14 and is a measure of the *hydrogen ion concentration* (acidity). Notice that low numbers indicate **high** acidity, i.e. high H^+ ion concentration (Fig I.7).

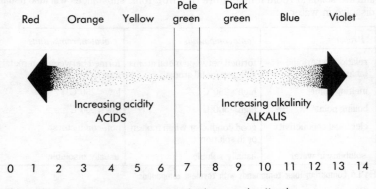

Fig I.7 Indicators: the colour changes of universal indicator on the pH scale

The colour of the indicator matches a number which indicates the acidity of the solution:

INDICATORS

- If a solution turns the Universal indicator *light green*, then it is neutral and has a pH of 7.
- If a solution turns the Universal indicator *yellow*, then it is a weak acid and has a pH of 6.

Other indicators can also be used to test for acidity or alkalinity:

Indicator	In acid	In alkali	In water (neutral)
litmus	red	blue	purple
phenolphthalein	colourless	pink	colourless

FOLLOWING AN ACID/ALKALI REACTION

Indicators can also be used to *follow the course of* a neutralization reaction between an acid and an alkali. If Universal indicator is added to a strong alkaline solution, the indicator will turn *violet*. If a solution of acid is added, a small amount at a time, then the indicator will change colour *through blue to green*, at which point the solution is *neutral* (i.e. the acid has reacted with all the alkali present). If acid continues to be added, the indicator will eventually turn *red*, indicating the solution is now strongly acidic.

Titration

You may have performed an experiment such as this and will have noticed how difficult it is to add just the right amount of acid to produce neutral solution (green colour with the indicator). If you need to produce a neutral solution, an accurate measuring device (a *burette*) is required and it would also be helpful to use an indicator (like phenophthalein) which has a more distinct colour change. The process is called **titration**. The acid is titrated into the alkali using an indicator to show when the exact amount of acid has been added (Fig I.8).

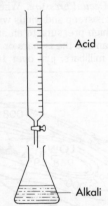

Fig I.8 Titration of acid and alkali

In the reaction between hydrochloric acid and sodium hydroxide a salt (sodium chloride) and water are produced. We can use the *titration method* to find the exact quantity of hydrochloric acid to exactly neutralize the sodium hydroxide solution:

hydrochloric acid	+	sodium hydroxide	→	sodium chloride	+	water
HCl(aq)	+	NaOH(aq)	→	NaCl(aq)	+	H$_2$O(l)

A measured volume of sodium hydroxide is placed in a flask, together with a few drops of indicator. The hydrochloric acid is then added from the burette, a little at a time, swirling the liquid in the flask to make sure the solutions are mixed. When the indicator *just changes colour*, then the exact amount of acid has been added to *neutralize* the alkali. The volume of the acid required can be read off the burette. If the same experiment is repeated, with the same volumes of acid and alkali, but without the indicator, then a neutral solution can be produced which contains only sodium chloride and water. This can be shown by evaporating the water in the flask, when crystals of sodium chloride will be left.

IRRADIATION

Irradiation is a method of preserving food by exposing it to radiation from radioactive chemicals. The radiation destroys all the bacteria and fungi so that the food keeps longer. Irradiated food is used for patients in hospitals who need sterile food. At the present time the law prevents irradiated food from being sold in shops to the general public as there is insufficient knowledge about the long-term effects of eating irradiated food.

ISOBARS

Isobars are the pressure contour lines on a weather map. The isobars join points of *equal pressure*. When the isobars are close together, the pressure gradient is steep and winds will be very strong. Wind generally moves from an area of high pressure to an area of low pressure.

You can see the isobars or pressure bars on the map in Figure I.9, labelled as 1000 millibars, just over 10N per centimetre square. ◀ Fronts ▶

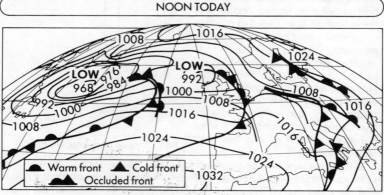

Fig I.9 Isobars: isobars link places which have equal pressure

ISOTOPES

The type of atom is determined by its **atomic number** (number of **protons**). Carbon is carbon because it has 6 protons. Chlorine is chlorine because it has

ISOTOPES

17 protons. It is possible, however, for atoms such as these to have different mass numbers. This means that they contain different numbers of **neutrons** in their nucleii. These atoms which are chemically the same, but differ in their mass numbers, are called Isotopes.

Chlorine has two isotopes: chlorine 35, with a mass number of 35, and chlorine 37, with a mass number of 37.

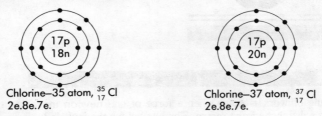

Chlorine—35 atom, $^{35}_{17}$Cl
2e.8e.7e.

Chlorine—37 atom, $^{37}_{17}$Cl
2e.8e.7e.

Fig I.10 Isotopes of chlorine

Both these atoms are identical chemically; the difference in their mass number is due to different numbers of neutrons in their nucleus (Fig I.10). In chlorine gas the proportion of these isotopes is always the same. There are 3 chlorine-35 atoms for every 1 chlorine-37 atom. The *average* atomic mass for chlorine is therefore:

$$\frac{35u + 35u + 35u + 37u}{4} = 35.5u$$

where u is the atomic mass unit.

Different elements have different proportions of isotopes (some of which may be radioactive). The **relative atomic mass** of an element is based on the average mass of all the atoms in the element and will not be a whole number.

JOULE

One joule of work is done when a force of one **newton** moves an object through a distance of one metre. The symbol for the joule is J.

KELVIN

This is the S.I. unit of temperature, symbol K. 1 K = 1°C. The *lower fixed point* is 273 K and the *upper fixed point* is 373 K.

KIDNEY

The kidneys are located in the abdomen (Fig K.1) and, when seen in section, consist of two separate layers – the outer, *cortex* and inner, *medulla*. Within the medulla are *nephrons*, where salts, such as urea, are passed from the blood into the **urine** (Fig K.2).

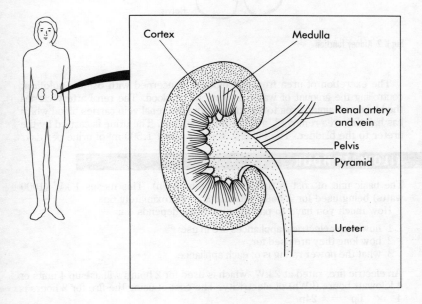

Fig K.1 Kidney structure

KIDNEY

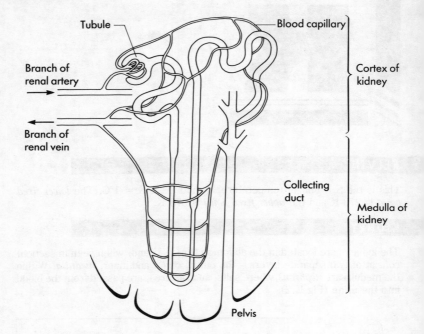

Fig K.2 Kidney function

The excretion of urea from the blood is concerned with **osmoregulation**, controlling the amount of water present in the blood. The renal artery carries the blood containing urea to the kidney, and the renal vein carries blood which has had the urea removed, away from the kidney. The urine is carried by the ureter to the bladder. The kidneys remove about 1,500 ml of urine per day.

KILOWATT HOUR

The basic unit of cost is the kilowatt hour (kWh). This means 1 kW (1000 watts) being used for 1 hour, which costs approximately 6p.

How much you have to pay for electricity depends on:

1 how many electrical appliances are in use;
2 how long they are used for;
3 what the power rating is of each appliance.

An electric fire, rated at 2 kW, which is used for 2 hours will use up 4 units or 4 kilowatt hours (kWh) of electricity. The cost of using the fire for 2 hours is 4 × 6p = 24p.

A table lamp, rated at 40 W and switched on for five hours, uses 0.04 × 5 units = 0.2 kWh. The cost of using the table lamp is 0.2 × 6p = 1.2p.

KINETIC THEORY

KINETIC ENERGY

Kinetic energy is the energy associated with *motion*. The kinetic energy, 'E', of a moving object depends on the mass (m) of the object and its velocity (v), as stated in the formula:

$$E = \tfrac{1}{2} m v^2$$

If a person of mass 60 kg is travelling with a velocity of 2 m/s then the kinetic energy = ½ × 60 × (2 m/s)² = 120 J

If the same person is travelling twice as fast at 4 m/s, then the kinetic energy = ½ × 60 × (4 m/s)² = 480 J

The kinetic energy has increased by four times, as the speed has doubled. If the mass of the person doubled to 120 kg, and the speed stayed at 2 m/s, the kinetic energy would be doubled,
E = ½ × 120 × (2m/s)² = 240J

These ideas about kinetic energy are very important when applied to real-life situations such as the braking distances of cars. When a car brakes to a stop, the kinetic energy is transferred to the brakes and the road. The amount of energy transferred is equal to the force of the brakes × the distance taken to stop. If one car has twice the mass of another car, but is travelling at the same speed, it will have twice the amount of energy and need twice the braking distance. If two cars have the same mass but one is travelling at twice the speed of the other car, it will have four times the amount of energy and need four times the braking distance.

KINETIC THEORY

All matter can be classified as either solid, liquid, or gas. These are called the three *states of matter*. The view that matter (solid, liquid or gas) is made up of particles which are in constant motion is called the kinetic theory. This idea can help us explain several properties of solids, liquids and gases (Fig K.3).

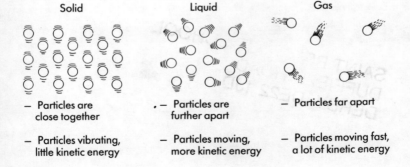

Fig K.3 Kinetic theory

KWASHIORKOR

The particles in a *gas* are moving very fast (they have a lot of kinetic energy) and are a great distance apart. The particles in a *liquid* are moving more slowly and are close together. In a *solid* the particles are very close together and are 'vibrating' rather than moving freely. This idea is often shown by a piece of equipment similar to that in Figure K.4. The motor turning very fast (providing a lot of energy) makes the metal spheres imitate a gas, but when it is moving more slowly (providing less energy) the spheres imitate a liquid.

◀ Brownian motion, diffusion ▶

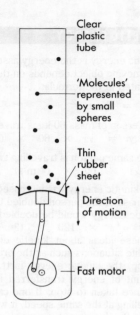

Fig K.4 Kinetic theory: this equipment imitates the movement of particles

KWASHIORKOR

Kwashiorkor is a deficiency disease which mostly affects young children, caused by lack of protein in the diet. The child appears to have a swollen body and does not grow properly. ◀ Balanced diet ▶

LASERS

The word laser means Light Amplification by Stimulated Emission of Radiation. Laser light is a very pure form of very bright light of a single concentrated colour which does not spread out in the way that ordinary visible light does. Laser light is very powerful and a beam of light with a power of 1 megawatt can be used to cut sheets of metal or to weld metal together. Lasers can also be used in hospitals by surgeons who are able to cut and seal without risk of infection. Laser surgery relies on **optical fibres**, which allow laser light to be transmitted to inaccessible parts of the body, such as a tumour inside the stomach. Two everyday uses of lasers are in compact disc players and bar-code readers at the checkout of a supermarket.

LATENT HEAT

This is the heat required to change the state of matter, for example from solid to a liquid. The temperature of the substance remains constant while it is taking in latent heat (Fig L.1). The *specific latent heat* is the amount of heat required to change the state of 1 kg of a substance, *without* a change of temperature.

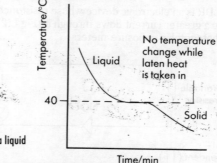

Fig L.1 Latent heat: cooling curve for a liquid changing to a solid at 40°C

LAVA

Lava is liquid magma which is forced out of a **volcano** during an eruption, and then cools to solid magma.

LEACHING

Water washes away (leaches) soluble mineral salts from the soil. As a result the soil is very poor with a poor yield of plants, so that **erosion** can easily occur.

LEAD POISONING

The symptoms of lead poisoning are many, but include headaches, stomach pains, constipation, loss of muscular control, and gout. Lead poisoning has also been linked to a loss of intelligence in young children. In severe cases a build up of lead in the body can lead to death. How much lead the body absorbs depends on the food we eat and where we live. Those people who eat large amounts of canned foods, live in old houses or near busy roads are most at risk. Lead is present in the solder in tin cans (silver solder is used in baby food cans). Some old houses still have old lead piping carrying water to the house, while living near busy roads means exposure to the lead which is emitted from car exhausts. Lead is added to petrol in the form of a lead compound (lead tetraethyl), to prevent the hot engine from 'pinking' – a condition where the petrol ignites before the spark is delivered, causing power loss and engine wear.

The body often only absorbs a small amount of the lead which passes through it, but what it retains can be stored in the bones. This is due to the chemical similarity between the Ca^{2+} and the Pb^{2+} ions.

LIEBIG CONDENSER

◀ Distillation ▶

LIGHT DEPENDENT RESISTOR (LDR)

An LDR is an electronic device whose **resistance** decreases when light falls on it, so a greater current flows through it (Fig L.2). A practical use of an LDR is in photographic exposure meters.

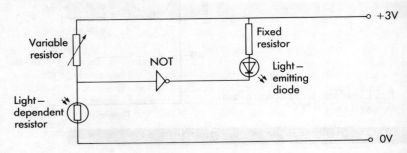

Fig L.2 **Light dependent resistor: a circuit activated by low light intensity**

LIGHT EMITTING DIODE (LED)

An LED is a small electronic device which emits red, green or yellow light when a current is flowing through it. A **resistor** is required in the circuit to limit the amount of current flowing to about 0.01 A. LEDs are used in electronic equipment such as calculators, measuring instruments, cash registers. The use of an LED as a warning light is shown in Figure L.2.

LIMESTONE

A **sedimentary** rock made of calcium carbonate. It is widely used for building, making cement, glass and lime.

LITMUS

Litmus is an **indicator** (a dye) which can detect the difference between an **acid** and an **alkali**. It can be used as a solution, or as litmus paper. Litmus paper consists of the solution soaked onto a type of blotting paper, which must be made wet before using. Litmus paper appears in two forms – red and blue, whereas the solution is a purple colour.

- When litmus turns **red**, acid is present.
- When litmus turns **blue**, alkali is present.

LIQUID

All matter can be classified as either *solid, liquid,* or *gas*. These are called the three *states of matter*. In a liquid, the particles are moving and are able to move around each other reasonably freely. Liquids take the shape of whatever they are poured into. Liquids at room temperatures usually consist of small molecules. The majority of solids (even those containing ions where the particles are strongly held together) will, when heated, melt to form liquids.

LOGIC GATES

These are electronic 'gates' in a circuit which only allow an output signal in response to particular input situations. They are termed *gates* because they are either open (logic 1) or closed (logic 0). A gate is a piece of electronic circuitry which is described by the way its output will become logic high = 1 when its input is changed. Once the function of a logic gate is known it can be built into a circuit with others and be of practical use in electronics.
Three common gates are the 'NOT', the 'AND' and the 'OR':

1. The NOT-gate is the simplest gate with one input and one output. It is sometimes called an *inverter* because it does just that, a high=1 input causes a low=0 output and vice versa. Its symbol and truth table are shown in Figure L.3.

LOGIC GATES

2 The AND-gate has two or more inputs and one output. The output goes high if both inputs are also high. A simple circuit as shown in Figure L.4a) shows that the lamp will only light when both switches are on. The symbol of the AND-gate and its truth table are shown in Figure L.4b). Any system that gives this output is called AND. Remember, the output is high if both one input AND the other is logic high.

3 The OR-gate is called an 'OR' because the output goes high if either of the inputs go high as shown in Figure L.5.

Input	Output
0	1
1	0

Fig L.3 NOT-gate symbol and truth table

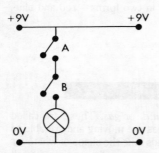

Fig L.4a) A simple AND-gate circuit

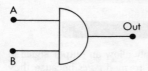

Input		Output
A	B	Z
0	0	0
1	0	0
0	1	0
1	1	1

Fig L.4b) AND-gate symbol and truth table

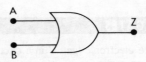

Input		Output
A	B	Z
0	0	0
1	0	1
0	1	1
1	1	1

Fig L.5 OR-gate symbol and truth table

Two other gates often used are the NAND and the NOR, as shown in Figure L.6.

Name of gate	Symbol	Truth table	Description
NAND	A, B → Out	0 0 1 0 1 1 1 0 1 1 1 0	Opposite of *AND* gate
NOR	A, B → Out	0 0 1 0 1 0 1 0 0 1 1 0	Opposite of *OR* gate. Output high if neither A *NOR* B is high

Fig L.6 NAND and NOR symbols and truth tables

LONGITUDINAL WAVES

These are waves where the particles of a medium move backwards and forwards, in the same direction as the wave motion (Fig L.7). Sound waves are longitudinal waves. ◄ Transverse wave ►

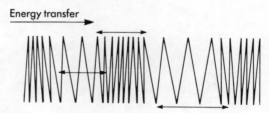

Fig L.7 Longitudinal waves: energy is being transferred along this longitudinal wave, but the particles are oscillating from left to right

LOUDSPEAKER

Figure L.8 shows the three main sections of the moving coil loudspeaker. When an alternating current is passed through the coil, it is pushed backwards and forwards, causing the paper cone to vibrate and give out sound waves. The frequency and amplitude of the alternating current which flows through the coil affects the type of sound produced.

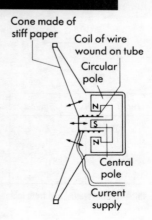

Fig L.8 Loudspeaker (moving coil)

LUNAR ECLIPSE

◄ Eclipse of the moon ►

LUNGS

◄ Breathing ►

LYMPHATIC SYSTEM

This is a system of fine tubes running throughout the body which removes excess tissue fluid from around the cells. This fluid (*lymph*) is returned to the blood system via a vein near the heart. *Lymph nodes* are situated along the lymph system to produce *lymphocytes* – white blood cells which produce antibodies to destroy bacteria. The lymph nodes also remove bacteria from the lymph.

MACROMOLECULES

◀ Molecules ▶

MAGMA

Magma is liquid rock at a temperature of about 1000°C. It cools to form **igneous rocks** such as granite.

MAGNESIUM

Magnesium (chemical symbol Mg) is a metallic element found in group 2 of the **periodic table**. It is silvery in colour and is very light for a metal. This latter property means that it is often used as part of an **alloy** in aircraft frames. Magnesium is near the top of the **reactivity series**, so it reacts readily. It will burn in air and react quickly with dilute acids. When magnesium reacts it forms the positive ion, Mg^{2+}. This ion is important to plants as a raw material for the manufacture of chlorophyll.

MAGNETIC CIRCUIT BREAKERS

Magnetic circuit breakers are sometimes used instead of **fuses**. These have the advantage that they are very easy to reset after they have broken the circuit.

MAGNETS AND MAGNETIC FIELDS

Magnets are solid objects which have a magnetic field around them. Magnets attract other magnetic metals such as iron and steel, but don't attract non-magnetic metals such as copper, tin, and zinc. Magnets are usually made of iron or steel or of magnetic alloys, and can either be *temporary* or *permanent* magnets. Temporary magnets are usually made of soft iron and lose their magnetism, whereas permanent magnets are usually made of steel, or a steel alloy.

The effect of a *magnetic field* around a magnet can be shown by using a plotting compass to find out which is the north-seeking or N pole of a magnet.

MAINS ELECTRICITY

The poles of a magnet are the ends of the magnets where the magnetism is strongest. If you place the plotting compass near the end of the magnet the needle of the compass is repelled from the N-seeking pole as shown in Figure M.1.

You can also show the magnetic field by shaking iron filings around a magnet. The iron filings line up along the lines of force and produce the patterns shown in Figure M.2a). These patterns show the line of magnetic force. The magnetic field patterns produced between two attracting poles and two repelling poles can also be seen using iron filings and are shown in Figure M.2b) and c).

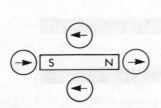

Fig M.1 Magnets: the compass needle is repelled from the N pole of the magnet

Fig M.2a) The magnetic field pattern around a magnet

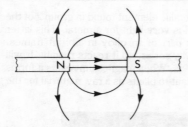

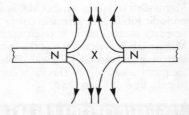

b) Attraction between unlike poles of two magnets

c) Repulsion between like poles of two magnets

When two magnetic fields come together there is either a force of attraction or a force of repulsion, and as a result there is a possibility of movement.

◀ Electromagnetism ▶

MAINS ELECTRICITY

Three key points to remember:

1. The voltage of the mains electricity in your home is 240 volts.
2. The direction of flow changes 50 times per second so its frequency is 50 hertz.
3. Live and neutral wires carry the mains electricity, and the insulation around the wires is colour coded so you know which is which.

◀ Three pin plug, Transmission of electricity ▶

MARK - RELEASE - RECAPTURE

This is a method of estimating the size of a **population**; for example, a population of snails. It involves using a special non-toxic paint or marker pen and following a set sequence:

1. Capture, count and mark a representative sample of a population.
2. Release the animals in the same area.
3. At a later stage, when the marked animals have mixed with the rest of the population, recapture and count the numbers of animals, and record how many of the marked animals are in the second sample.
4. Use the formula below to estimate the total population:

$$\frac{\text{number in first sample} \times \text{number in second sample}}{\text{number of marked animals recaptured}}$$

MASS

The mass of an object is the amount of a material measured in kilograms. This amount does not change. An astronaut landing on the Moon has the same mass as on Earth, but only one sixth of the **weight** (Fig M.3).

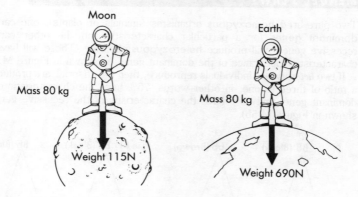

Fig M.3 Mass: comparative weight on Earth and on the moon

MASS NUMBER

The total number of **protons** and **neutrons** contained in the **nucleus** is called the mass number (each proton and neutron has a **mass** of 1u). There are usually about the same number of protons as neutrons in a nucleus.

You can work out the structure of an **atom** from two numbers:

- Atomic number = number of *protons* (= number of *electrons*)
- Mass number = number of *protons* + number of *neutrons*

◀ Atomic structure ▶

MASS SPECTROMETER

◄ Atomic mass ►

MEIOSIS

Meiosis is the process of cell division which takes place in the sex organs to form the **gametes** or sex cells. In the male human it occurs in the testis to form sperm, the male sex cell. In the female human it takes place in the ovary to form ova (egg cells) the female sex cell.

During meiosis the chromosome number is reduced to the *haploid* (half) number of **chromosomes**, and four non-identical cells are produced. Meiosis is sometimes described as reduction division (Fig M.4).

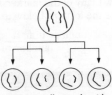

4 new cells, each with only 2 chromosomes

Fig M.4 Meiosis in a cell with four chromosomes

MENDEL'S LAWS

Two pure-bred **homozygous** organisms (animals or plants), one carrying **dominant genes** for a particular characteristic and the other carrying recessive genes, will produce **heterozygous** offspring. These will have the characteristic appearance of the dominant gene, as shown in Figure M.5a).

If two *heterozygous* individuals reproduce, then the offspring are produced in a ratio of three to one; in other words, 75% have the characteristic of the dominant gene, and 25% have the characteristic of the recessive gene, as shown in Figure M.5b).

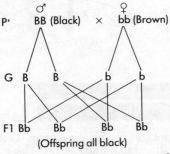

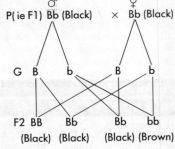

♂ = male
B = dominant (black) allele
b = recessive (brown) allele
P = parents

♀ = female
× = crossing (mating)
G = gametes
F1 = offspring in first generation ('first filial')

a) Homozygous parents

b) Heterozygous parents

Fig M.5 Mendel's Laws

MENSTRUAL CYCLE

This is a periodic change which occurs in a woman's body about every 28 days (Figure M.6). During the first few days of the cycle the extra lining of the *uterus* breaks down and is released from the body. During the next 10 days an egg ripens in the ovary and the uterus lining thickens again. On about the 14th day the egg is released from the ovary during *ovulation*, and this is when **fertilisation** can occur. If no fertilisation occurs, then the uterus lining breaks down and is released on about the 28th day. ◄ **Menstruation** ►

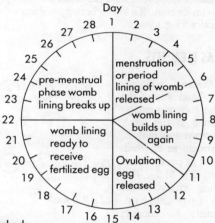

Fig M.6 The Menstrual cycle

MENSTRUATION

Menstruation is the passing from a woman's body of the extra lining of the *uterus* (womb) which has developed. This extra thick lining is developed to receive the embryo after **fertilisation**.

If fertilisation has not taken place, the lining breaks down and is released as blood and mucus through the *vagina*. In women this happens about every 28 days, but stops during pregnancy.

METABOLISM

This process covers all the chemical reactions which take place in an organism. The reactions are usually controlled by **enzymes**.

For example, the release of energy from sugar is a metabolic process whose waste products need to be removed by **excretion**.

METAL

All metals are **elements** (an individual metal contains only one type of atom). Metals exist as **compounds** in rocks; those rocks which contain large quantities of metals are called *ores*. Metals can be extracted from ores by

METAL

techniques such as **smelting** (reduction using carbon) or **electrolysis**. Metals have a number of useful properties:

- conductors of heat and electricity
- solids (except mercury)
- strong, malleable and ductile
- high melting points and densities
- shiny

There are exceptions to these general properties and different metals exhibit these properties to varying degrees, e.g. copper is used as electrical wiring because it is a better conductor than many. It is important to realize, however, that the choice of a metal for a particular job not only depends on its properties, but also on its cost. (Gold is a better conductor than copper but is not used in electrical wiring.)

▶ METALS AS ELEMENTS

All metals are *elements* and are grouped on the left hand side of the **periodic table**. The common everyday metals are found in the block referred to as the **transition metals**. Metals are **giant structures** in which the metal atoms exist as positive ions in a sea of electrons (Fig M.7). This gives rise to their characteristic physical properties of strength, hardness and the ability to conduct electricity.

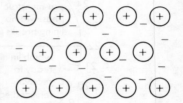

Fig M.7 Metals

In many of their reactions metal atoms form positive ions, the charge on the ion depending on the metal's *position* in the periodic table (Fig M.8).

Group	Charge on ion	Formula	Metal
1	1+	Na^+	sodium
2	2+	Ca^{2+}	calcium
3	3+	Al^{3+}	aluminium

The transition metals can form ions with different charges; they are said to have *variable valency*. In *compounds*, the charge on the metal ion is indicated by Roman Numerals. *Transition metal compounds* are often coloured.

Compound	Colour	Formula	Metal ion
copper(II) sulphate	blue	$CuSO_4$	Cu^{2+}
iron(II) sulphate	green	$FeSO_4$	Fe^{2+}
iron(III) oxide	red	Fe_2O_3	Fe^{3+}
copper(I) oxide	red	Cu_2O	Cu^+
copper(II) oxide	black	CuO	Cu^+

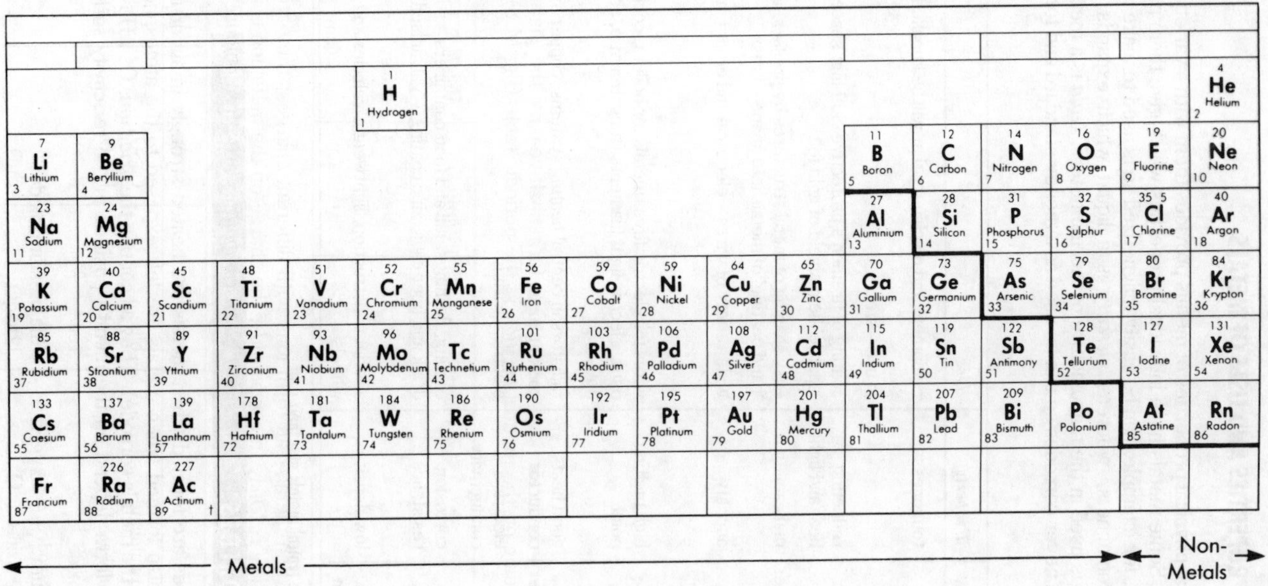

Fig M.8 The position of metals in the periodic table

PROPERTIES AND USES OF METALS

The general physical properties of metals provides them with a variety of uses (Fig M.9). Some metals have individual properties which give them particular uses. *Iron*, for example, can be easily magnetized so is used to make magnets and electromagnets, whereas *mercury* is a liquid which expands well on heating so is used in the manufacture of thermometers. *Gold* is a metal which is rare and never corrodes so can be used as a money standard and for making jewellery.

Metal	Property	Use
Iron	toughness	equipment and machinery that will be knocked about in use – steel
Aluminium, gold	reflecting heat and light	coating of firemen's protective clothing; space 'shuttle' heat shield of gold foil
Zinc, copper	malleability	easy shaping of metal structures by presses, also 'hand beating' of metals into shape – brass
Copper, aluminium, gold	ductility	wide variety of wires – electrical and ornamental
Iron, aluminium, tungsten	high melting point	wires for electric fires, metals for boilers, cookers, pans, electric light filaments (tungsten m.p. about 3500°C)
Aluminium, iron, copper	good heat conductor	radiators in central heating systems, copper for cooking pans
Copper, aluminium	electrical conductivity	electrical wiring
Lead, aluminium, iron, zinc	corrosion resistance	roofing, flashings, foil and food containers – soft drink and beer cans, zinc coating – 'galvanised' steel
Aluminium, magnesium	low density	aircraft construction, lightweight vehicles and wheels

Fig M.9 Some typical metals and their uses

METAL OXIDES

Metal oxides are formed when metals are heated strongly in air or oxygen. They are also formed when metals are left exposed to the atmosphere and rain (when the metals corrode). They all contain the *oxide ion*, O^{2-}. How easily the oxides form depend on the metal's position in the **reactivity series**. For example:

magnesium + oxygen → magnesium oxide
$2Mg + O_2 → 2MgO$

All metal oxides are **bases** and react with **acids**, neutralizing them:

magnesium + hydrochloric acid → magnesium + hydrogen
oxide   chloride

$MgO + 2HCl \rightarrow MgCl_2 + H_2$

Most oxides are not soluble in water; those that are soluble react with the water to form hydroxides and are called **alkalis**. For example, the oxides of sodium, potassium and calcium will react with water to form hydroxides.

METAMORPHIC ROCKS

Metamorphic rocks are formed from the action of pressure and heat on both **igneous** and **sedimentary rocks**. Movements of the **Earth's crust** break up sedimentary rocks and push them down into the hotter areas of the Earth where metamorphic rock is formed. For example, limestone, a sedimentary rock is changed into *marble* by intense heat and pressure; compressed *mud* can be changed into slate.

METHANE

Methane is a colourless, odourless gas which burns, producing a lot of heat. It is found in oilfields and bubbles up through swamps, marshes and rubbish tips – in fact any area where dead material is rotting down. Natural gas from the North Sea is methane (for safety reasons a smell is *added* to the methane so that people can detect leaks, etc).

Methane is a hydrocarbon and has the chemical formula CH_4. When methane burns it produces *carbon dioxide* and *water* :

$$CH_4 + O_2 \rightarrow CO_2 + 2H_2O$$

If, however, methane burns in a *limited* supply of air, *carbon monoxide* will be produced which is highly poisonous.

MICRO-ORGANISM

Micro-organisms (microbes) are **bacteria, fungi** and **viruses**.

Viruses are very small simple structures which lack a nucleus, cytoplasm and cell membrane. They are very infectious and cause many diseases such as influenza, polio, etc.

Fungi are important in the decomposition of dead organisms, and are also important in the food industry, for example in baking and brewing.

◄ Bacteria, nutrient cycles ►

MICROPHONE

A microphone is a device for converting a pattern of sounds into electrical impulses. In the microphone is a thin metal sheet called a *diaphragm* which vibrates when sound waves hit it. These vibrations push against carbon granules in the microphone and alters the **resistance** of the granules. A variable current is then passed to the receiver which changes the electrical input into sound waves. ◄ Loudspeaker, tape-recording ►

MICROWAVES

Microwaves are radio waves which have a very short wavelength. They are used for **radar** and telephone and television links on an international basis, via geostationary **satellites**.

One common use of microwaves is in cooking food. The microwaves cook the food by heating the *water molecules* in the food, and also destroy **bacteria**.

MINERALS (SALTS)

Minerals are important substances in a **balanced diet** as they are needed for the formation of complex **molecules** in the body.

Mineral	Needed for:	Deficiency causes:	Source:
Phosphorus Calcium	bones and teeth formation	brittle bones and teeth	milk cheese fruit
Iron	Haemoglobin formation	anaemia (insufficient haemoglobin)	liver eggs spinach
Iodine	Growth hormone (thyroxin) formation	goitre (swollen thyroid glands)	seafood salt

MITOSIS

Mitosis is the process of cell division which takes place in all the body cells except the sex cells. Each **chromosome** in the nucleus of the cell replicates to form two chromosomes. The nucleus then divides into two and two identical daughter cells are produced containing the full or haploid number of chromosomes (Fig M.10). ◄ **Gametes** ►

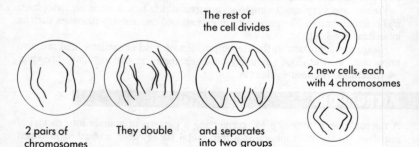

Fig M.10 Mitosis in cell with four chromosomes

MIXTURE

Mixtures are substances which contain various amounts of **elements** and/or **compounds** mixed together. They can easily be physically separated. For example, a mixture of iron and sulphur can easily be separated with a magnet.

MOLAR SOLUTIONS

Chemical reactions often take place in *solution*, so if we need information about the numbers of particles present we need to know the concentrations of the reactants in solutions.

Concentrations are given in mol/dm^3; a solution which contains 1 mol/dm^3 contains 1 mole per dm^3 (litre). This is often expressed as a 1M solution (1 molar). Thus a 2M solution contains 2 mol/dm^3. Therefore, a 1M solution of sulphuric acid contains 98g of H_2SO_4/dm^3.

We can use this information to work out the number of moles present in a solution. For example, how many moles are present in 25cm^3 of a 2 mol/dm^3 solution of NaOH?

No of moles = concentration × volume (in dm^3)
 = 2 × 25/1000
 = 0.05 moles.

Note: 25/1000 converts the volume from cm^3 to dm^3.

We can use a similar method to calculate the *concentration* of solutions. For example, 25cm^3 of a 2 mol/dm^3 solution of sodium hydroxide exactly reacts with 10cm^3 of sulphuric acid. What is the concentration of the sulphuric acid? First, look at the equation of the reaction:

$$2NaOH + H_2SO_4 \rightarrow Na_2SO_4 + 2H_2O$$

Now, from the question, the number of moles of NaOH used = concentration × volume
 = 2 × 25/1000 = 0.05 mole

From the equation, 2 moles of NaOH reacts with 1 mole of H_2SO_4; so 0.05 mole of NaOH reacts with 0.025 mole of H_2SO_4.

Then, using the information that no. of moles = concentration × volume:
0.025 = concentration × 10/1000
2.5 = concentration

The concentration of H_2SO_4 = 2.5 mol/dm^3.

MOLE

There are two facts that apply to any chemical change:

- The total mass of the products = the total mass of the reactants
- The total number of atoms in the products = the total number of atoms in the reactants.

These two facts allow us to make calculations involving chemical reactions. To do this satisfactorily, we need a method of counting the number of particles present. The concept of the *mole* was developed for this purpose and is a measure of the amount of substance. It is equivalent to 6×10^{23} particles.

When large amounts of coins are handed into the bank they do not count them individually but instead they weigh them. In order to convert the weight of the coins to a number, the bank needs to know certain facts:

- How much does 100 1p coins weigh?
- How much does 100 2p coins weigh? etc.

We can apply the same principle to particles, since each atom has its own distinct mass: *relative atomic mass*. However, the number we use – the *mole* – has to be very much larger than 100, since the mass of each atom is very small. Converting numbers to mass is very simple.

Moles of atoms

The **relative atomic mass** of an **atom** in grams, contains 1 mole of atoms:

Atom	Atomic mass	Mass of 1 molecule of atoms
hydrogen	1	1g
carbon	12	12g
oxygen	16	16g
chlorine	35.5	35.5g
sodium	23	23g

So in 32g of *oxygen* we have 2 moles of atoms, and in 8g of *oxygen* we have 0.5 mole of atoms.

Moles of molecules

The *molecular mass* in grams contains 1 mole of molecules:

Molecule	Molecular formula	Mass of molecule	Mass of 1 mole of molecules
oxygen	O_2	32	32g
water	H_2O	18	18g
carbon dioxide	CO_2	44	44g
hydrogen	H_2	2	2g

Note: the mass of a molecule is found by adding the individual atomic masses.

For example, for water (H_2O), the mass of the molecule = $2\times H + O = 2\times 1 + 16 = 18g$.

So in 36g of *water* there are 2 moles of molecules, while in 16g of *oxygen* there is 0.5 mole of molecules.

Moles of ionic compounds and ions

The same rules apply for formulae representing *ionic compounds* and for individual ions:

Compound or ion	Formula/ion	Formula mass/ mass of ion	Mass of 1 mole
hydrochloric acid	HCl	36.5	36.5g
sodium hydroxide	NaOH	40	40g
sulphuric acid	H_2SO_4	98	98g
sulphate ion	SO_4^{2-}	96	96g
chloride ion	Cl^-	35.5	35.5g

Using the mole idea

What mass of magnesium chloride is produced when 12g of magnesium reacts with excess hydrochloric acid? Note that 'excess' means that you have more acid than you need, so the mass of this compound is not a restriction and can be ignored.

From the equation of the reaction:

$$Mg + 2HCl \rightarrow MgCl_2 + H_2$$

We can see that *one* mole of magnesium produces *one* mole of magnesium chloride. To calculate the *amount* of magnesium chloride, we must first convert masses to moles. Since the atomic mass of magnesium is 24, the mass of 1 mole is 24g. Therefore the number of moles in 12g of Mg = 0.5.

From the equation, 1 mole of Mg produces 1 mole of $MgCl_2$, so 0.5 mole of Mg produces 0.5 mole of $MgCl_2$.

Then, we convert moles to mass. If 1 mole of $MgCl_2$ has a mass of 95g, then 0.5 mole has a mass of $1/2 \times 95g = 47.5g$.

Answer : 47.5g of $MgCl_2$ is produced.

MOLECULAR MASS

◀ Mole ▶

MOLECULE

All matter is made up of **particles**; there are three different types: **atom**, **molecule** and **ion**. The molecule is a particle which contains two or more atoms chemically joined together (Fig M.11). Molecules can contain the same type of atom, or different atoms chemically joined (bonded). Each molecule has a name and a chemical formula to represent which atoms are joined together.

MOLECULE

For example:

Molecule	Name	Formula
H – H	hydrogen	H_2
H\O/H (H–O–H)	water	H_2O

The numbers show the proportions of each atom present. They refer to the atoms immediately *before* the number, which is always written below the line (subscript). For example, glucose (a sugar) – $C_6H_{12}O_6$ – contains 6 atoms of carbon, 12 atoms of hydrogen and 6 atoms of oxygen.

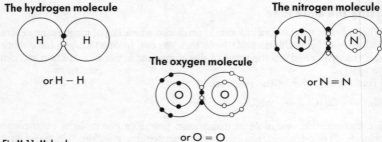

Fig M.11 Molecule

MACROMOLECULES

Molecules are usually small particles containing only a few atoms, such as those found in gases and liquids. Some molecules, however, are very large, containing thousands to millions of atoms – these are called macromolecules.

Elements forming macromolecules

Carbon forms large numbers of covalent bonds between its atoms. It can do this is two ways to form diamond or graphite. These two forms of carbon are called **allotropes** (Fig M.12). Some other elements can form allotropes, but these are not necessarily giant molecules (e.g. sulphur).

Fig M.12 Macromolecule: allotropes of carbon

Diamond is very strong because each carbon atom is linked to four others. A diamond crystal is one giant molecule. Graphite is very strongly bonded in layers; each layer is a giant molecule, but the forces holding the layers together are weak so they slide over each other. This property is made use of in pencil 'lead', which is really graphite.

Silicon is in the same group as carbon; it too has similar abilities to form giant structures. The structure of silicon is the same as that of diamond. Silicon dioxide, a compound of silicon, has a giant structure, the atoms being bonded by covalent bonds. We come across this substance quite often, it appears as sand, as quartz in rocks and can be made into glass.

Macromolecular compounds

Many macromolecules are based on carbon and its ability to form long chains. They have a number of useful properties, often existing in living things e.g. wool (a **protein**), starch (a carbohydrate) or are man-made from carbon-based compound, for example detergents and **polymers**. Scientists can synthesise (make) macromolecules such as polymers to have specific properties. For example, 'polythene' is a long carbon chain to which is attached hydrogen atoms; by replacing the hydrogen atoms with fluorine atoms a different polymer – PTFE – can be made. This is used on non-stick frying pans. The properties of these *compounds* are determined by the way the carbon atoms are joined in the chain (single bonds or double bonds etc.) and the type of atom attached to the chain.

Properties of macromolecules

Macromolecules have very high melting points and boiling points. They are not soluble in water and will not generally conduct electricity (except graphite or silicon, which is a weak conductor; it is regarded as a 'semi-conductor'. (This property makes silicon valuable for use as 'microchips' in computer technology.)

The large molecules (polymers), such as starch, have higher melting points and boiling points than ordinary molecules, but not as high as giant molecules or ionic compounds (which are also **giant structures**). They are not very soluble in water (many not at all) and will not conduct electricity.

MOMENTUM

The momentum of an object is a product of its mass and its **velocity**. For example, an object with a mass of 10 kg moving with a velocity of 5 m/s has a momentum of 50 kg m/s.

MONOHYBRID CROSS

◄ Mendel's Laws ►

MONOMER

Polymers which are long chain molecules (**macromolecules**) can be made by combining together much smaller molecules called monomer units (Fig M.13).

MOON

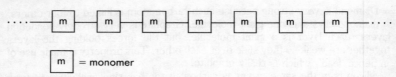

Fig M.13 Monomers: these monomer units which form the polymer can be the same or maybe a mixture of different types

Plants build up starch (a polymer) by combining glucose molecules (monomers) together. The plant manufactures glucose by **photosynthesis**. Many man-made polymers have been produced, for example:

Monomer	ethene	chloroethene	tetrafluoroethene
Polymer	polyethene (polythene)	PVC	PTFE

MOON

The Moon is a **satellite** which takes 28 days (a lunar month) to orbit the **Earth**. The Moon is held in its orbit by gravitational attraction between it and the Earth and the distance between the Earth and the Moon is approximately 384 to 400 km. The moon also rotates on its own axis every 28 days so the same side of the Moon faces Earth all the time. The Moon has no atmosphere and no water.

Figure M.14 shows how the different phases of the Moon appear when the Moon is viewed from Earth. When the Earth is between the Sun and the Moon we can see all the light from the Sun which is reflected by the Moon, and the Moon therefore appears to be a full Moon. ◄ **Eclipses** ►

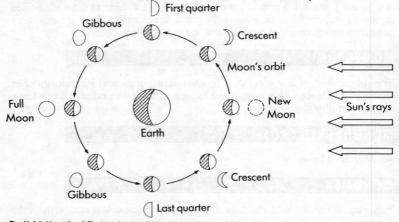

Fig M.14 How the different phases of the Moon are caused

MOTORS

Motors are devices which transfer electrical energy to **kinetic energy**, in other words, they produce motion.

SIMPLE MOTORS

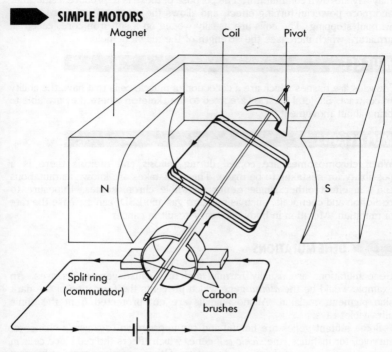

Fig M.15 The construction of a simple electric motor

A simple motor contains several coils of wire, wound on a core, which is pivoted on an axle between two permanent magnets, as shown in Figure M.15. When an electric current is passed through the coil the armature turns and produces mechanical motion. The coil is connected to a power supply by two carbon contacts called brushes. These are held in position against two halves of the **commutator** which is a split ring made of copper. When a direct current is passed through the coil, the magnetic field created is attracted to the opposite poles of the permanent magnets and this causes the coil to spin in a clockwise direction. When the N and S poles of the coil lie opposite the S and N poles of the permanent magnets the coil should stop turning, but it carries on spinning because the two brushes now press against the opposite half rings of the commutator. The current now flows in the opposite direction and this results in the N and S poles of the coil being reversed. The coil then spins round to the S and N poles of the permanent magnets, and once again the current direction is reversed as the coil is about to stop.

MORE COMPLEX MOTORS

Real motors which are used in everyday appliances such as electric drills, washing machines and food mixers usually have several coils, each of which may have its own commutator. The purpose of these is to produce a smoother and more powerful turning effect, and allows the motor to run more evenly without stopping. The coils are usually wound on a soft iron core, called an armature which increases the strength of the magnetic field.

MUSCLE

Muscles are tissues which are a collection of muscle cells that have the ability to contract. Skeletal muscles are fixed to the **skeleton** where they are able to bring about movement.

MUTATIONS

When **chromosomes** are copied during **mitosis** and **meiosis** there is a possibility for mistakes to be made. These mistakes are known as mutations and can effect either single **genes** or whole chromosomes. Exposure to **radiation** and chemicals, such as mustard gas and LSD, can increase the rate of mutation. Mutation in body cells may result in cancer.

GENE MUTATIONS

Gene mutations are usually harmful and may cause genetic diseases. An example would be the *albino* gene which prevents the formation of the dark skin pigment, melanin. Albino animals are not protected from the sun's ultra-violet rays.

Some mutant genes are helpful and can improve an organism's chance of survival, for instance, the *sickle-cell* gene, which affects the red blood cells in humans, can give some immunity to malaria. Other mutations seem to have no effect and are termed neutral – they appear not to affect survival. The sickle-cell gene can be harmful if a child gets it from both parents. This is one of some three thousand known genetic diseases and we now have genetic counsellors to help affected people. About 5% of children admitted to UK hospitals are suffering from genetic diseases.

CHROMOSOME MUTATIONS

Chromosome mutations occur when chromosomes are altered during meiosis; bits may be broken off or added to chromosomes and sometimes whole chromosomes may be lost or gained. The faulty gametes which are produced may be fertilized and produce zygotes with damaged chromosomes, or too few or too many chromosomes. An extra number 21 chromosome in humans produces a *Down's Syndrome* child with a low mental age and very characteristic facial features which led to the previous name for this genetic

MUTATIONS

disorder – mongolism. You may indeed know of a Down's person who has developed useful skills through special training and is a happy member of a family. Such diseases can now be detected in the early stages of pregnancy by taking a sample of fluid from the *amniotic fluid* in the womb and examining some of the embryo's cells.

NAND-GATE

◀ Logic gates ▶

NATURAL SELECTION

Natural selection is the theory we use to explain evolution. Darwin's voyage around the world in the 1830s aboard HMS Beagle provided the key for him to develop and refine his theory in the years that followed.
Darwin's observations were that:

- organisms produce large numbers of offspring;
- the offspring vary considerably;
- many offspring die before adulthood;
- many offspring do not survive to breed;
- they die because they can't overcome problems – starvation, being eaten by predators, being fatally injured, suffering from disease etc.

Darwin described this as a struggle for survival against a harsh environment. Scientists today call the difficulties 'selection pressures'. It is these pressures that determine which individuals survive. Those best adapted survive to pass their **genes** on to the next generation. Hares that can run fastest will escape the fox and so genes for powerful leg muscles will be 'selected' and over many generations the performance of the species will be enhanced.

If the environment changes, the process of natural selection allows the **species** to adapt to the new situation, the most advantageous variations surviving to breed. Without **variation** a species is very likely to become extinct.

The Peppered Moth is a good example of Natural Selection in that the colour of the moths has changed in recent times. The darker, mutant moths became more common in industrial areas when the Industrial Revolution blackened and polluted the environment. The dark moths were still rare in the countryside which was unpolluted and where the lighter form of moth was dominant (Fig N.1). Camouflage is the key to understanding this phenomenon; **predators** could easily find light-coloured moths on soot covered bark around cities, and so selection favoured the dark genes. In the countryside the reverse was true. Now that industrial pollution is less severe the situation should change again with the light-coloured form becoming dominant both in cities and in the countryside.

Fig N.1 Natural selection

	Percentage of each form	
Year	Dark	Pale
1848	1	99
1894	99	1

NEAP TIDE

◀ Tides ▶

NEGATIVE FEEDBACK

◀ Feedback systems ▶

NEUTRALIZATION

A neutral solution is one which has a pH of 7; for example, pure water. Neutralization is a chemical reaction which involves an **acid** and a **base**, the result of which is a **salt**. The base in the reaction could be either a metal oxide, a metal hydroxide, or a carbonate. Examples of these reactions are as follows:

hydrochloric acid + magnesium oxide → magnesium chloride + water

hydrochloric acid + sodium hydroxide → sodium chloride + water

hydrochloric acid + calcium carbonate → calcium chloride + water + carbon dioxide

In each case both the acid and base have been neutralized.

NEUTRON

The neutron is a subatomic particle found in the nucleus of an **atom**. It has a mass unit of 1u and no charge. There is no set rule for determining the number of neutrons in an atom as there is for **protons** and **electrons**, but a good rule of thumb is that there are about the same number of neutrons as protons. Some atoms, for example chlorine atoms, can have different numbers of neutrons in their nuclei; these are known as **isotopes**. It is fast-moving neutrons that cause chain reactions in nuclear reactors.

NEWTON

This is the unit of measurement for a force; symbol N. A force of 1 newton gives an acceleration of 1 m/s^2 to a mass of 1 kg.

NEWTON'S LAWS

Sir Issac Newton discovered three basic *laws of motion*:

1. An object will remain at rest or will continue to move at the same speed in a straight line unless it is acted on by a force. This means that once an object is moving it will keep on moving unless something stops it or changes its motion. For example, the force of **friction** will slow down the rate at which objects move. An aircraft moves by the force of the engines pushing the aircraft forwards which is equal to the force of friction which creates a drag effect. These two forces balance each other so the aircraft moves at constant speed.
2. The rate of change of momentum is equal to the applied force. This means that how much an object accelerates or decelerates depends on the size of the force acting on the object (Fig N.2). This relationship is expressed as

    force = mass × acceleration

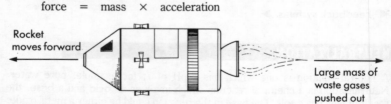

Fig N.2 Newton's Laws: the rocket accelerates due to the force of the gases being pushed out

3. Action and reaction are equal and opposite. This means that when a force acts, an equal force acts in the *opposite* direction. For example, gravity is exerting a force on you to pull you *down*, but your muscles are exerting an upward force to make you stand *up*.

NITRATES

Nitrates are chemical compounds (salts) which contain the nitrate ion, NO_3^-, for example copper nitrate, $Cu(NO_3)_2$. All nitrates are soluble in water, and

NITROGEN CYCLE

are used in many manufacturing processes, especially of fertilizers and explosives. Consequently, nitrate production is very important.

MANUFACTURE OF NITRATES

Nitrate manufacture starts with the Haber process, in which hydrogen and nitrogen from the air are combined to produce *ammonia*:

$$N_2 + 3H_2 \underset{\text{heat}}{\overset{\text{catalyst}}{\rightleftharpoons}} 2NH_3$$

The ammonia is then oxidized to produce nitric acid, from which a variety of nitrates are obtained.

Nitrates as fertilizers

Nitrates are important as fertilizers because plants need nitrogen to build **proteins**. Plants take in nitrates from the soil though their roots. A build-up of nitrates in river water has been blamed on farmers using nitrate fertilizers instead of using more traditional methods such as organic fertilizers (manure) from cows, pig manure etc. Scientists have shown that the increased rise in nitrate levels is not only due to use of fertilizers, but also to changes in farming practice. When land is left bare in winter, the rain washes out the natural nitrates in the soil, as well as nitrates applied by farmers.

Nitrates in drinking water

Too much nitrate in the drinking water can not only make babies ill, but has also been linked to increases in stomach cancer (although there is no firm evidence for this). A major problem of nitrates in the rivers and lakes is that they cause increased growth of water plants and algae, which clog waterways. When these plants die they are decomposed by bacteria, which in doing so use up the oxygen in the water. The fish are deprived of oxygen and subsequently die. ◄ Nitrogen cycle ►

NITROGEN

Nitrogen is a colourless, odourless gas that makes up about 80% of the air around us. Nitrogen gas contains nitrogen molecules (N_2). Nitrogen itself is not very reactive, but it is needed by plants in the form of nitrates (NO_3^-) to grow. ◄ Nitrogen cycle ►

NITROGEN CYCLE

Plants such as peas, beans and clover, are able to absorb nitrogen gas from the air through special swellings on their roots called *nodules* (Fig N.3). These nodules contain nitrogen-fixing bacteria which take in or 'fix' the nitrogen as **nitrates**. The nitrates are then used by plants to make **proteins**. The proteins are taken in by animals when they eat the plants, and are returned to the soil when animals and plants are decomposed by **bacteria** and **fungi** which live in soil. The **decomposers** form ammonium compounds, which are converted into nitrates by nitrifying bacteria.

Farmers often plant peas, beans or clover to help increase the amount of nitrates in the soil, instead of adding nitrogen in the form of nitrate fertilizers. The peas, beans and clover can then be ploughed back into the soil, and the nitrates can be used by other plants to make proteins. Some nitrates are lost from the soil when denitrifying bacteria convert the nitrates into nitrogen gas, which is released into the air. However, some nitrates are added to the soil when lightning converts nitrogen into nitrates.

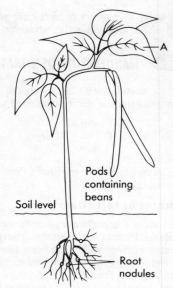

Fig N.3 Nitrogen cycle: a bean plant with root nodules containing nitrogen-fixing bacteria

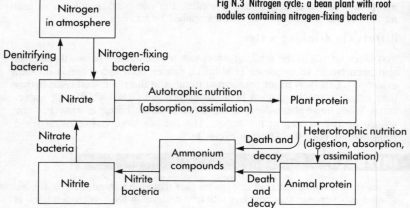

Fig N.4 Nitrogen cycle

NOBLE GASES (INERT GASES)

The noble gases are a family of **Elements** in the **Periodic table** (Group 0), sometimes referred to as the *inert gases* (Fig N.5). These gases show no chemical reactivity (with a few exceptions) because they have stable filled outer electron shells or orbitals. They do not form molecules but exist as separate atoms. The gases have various uses, depending on their inert behaviour.

For example:

- Argon is used to fill light bulbs – inert argon does not react with a hot-wire filament in the way that oxygen would.
- Helium is used in airships – it is lighter than air but is not inflammable, unlike hydrogen.
- Neon is used in street-lamps and gives a characteristic pale blue glow.

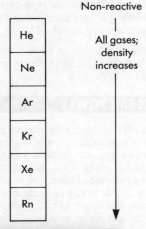

Fig N.5 The noble gases – Group 0

NOT-GATE

◀ Logic gates ▶

NUCLEAR FISSION

The atoms of radioactive material have unstable nuclei that break down releasing energy, either as radiation (**gamma rays**) or as kinetic energy from **alpha** and **beta** particles. Nuclear fission is a process in which a radioactive nucleus splits into fragments; this occurs naturally in some elements which have very large unstable nuclei. When this happens often a few **neutrons** are released as well.

The nucleus of a rare form of uranium (uranium-235) will break down, at the same time releasing a few neutrons. These fast-moving neutrons can then strike another nucleus. When this happens the second nucleus will also immediately break down, releasing yet more neutrons. Each fission (breakdown) produces more neutrons, which in turn cause other breakdowns. This is called a *chain reaction*, and can happen very quickly. Each time a nucleus breaks down, a large amount of energy is released; this results in a rapid rise in temperature of the uranium and its surroundings. This type of reaction takes place in a nuclear reactor, which is fed with concentrated uranium-235. ◀ Nuclear power ▶

NUCLEAR FUSION

Nuclear fusion is the opposite of **nuclear fission**. In nuclear fusion energy is released when small nuclei are *joined together* to form larger nuclei. This process is happening all the time in the Sun, and is the source of the Sun's energy. Scientists and technologists have been trying to produce nuclear fusion reactors on Earth for a long time but have still not overcome two major problems:

1 bringing the particles together fast enough;
2 building a 'container' for the reaction that can withstand the high temperatures involved.

The main advantage of nuclear fusion over nuclear fission is that it produces far less radioactive by-products and could use readily available *deuterium* (an isotope of hydrogen).

NUCLEAR POWER

The source of energy for nuclear power comes from the energy stored in the nuclei of a particular type of uranium, uranium-235, which has 92 protons and 143 neutrons in the nucleus. In a nuclear power station, the uranium is in the form of fuel elements in the reactor core.

During **nuclear fission**, the uranium nuclei are hit by slow-moving neutrons and the uranium nucleus splits into two smaller parts, giving out energy in the process. The neutrons which are released from the uranium nucleus are then used to split more uranium-235 nuclei in a chain reaction. In the nuclear reactor, this reaction is controlled by control rods, made of boron, which absorb neutrons. The heat produced by the reaction is carried away by a coolant liquid to a *heat exchanger* where it is used to generate steam which drives turbines to generate electricity (Fig N.6).

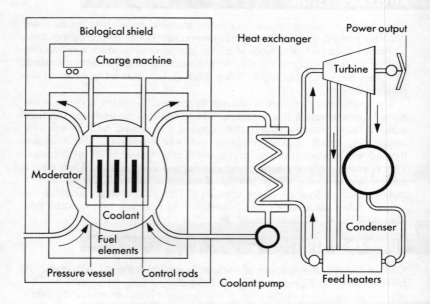

Fig N.6 Nuclear power: a simplified diagram of a nuclear power station

Nuclear power is an important source of energy as an alternative to fossil fuels. One of its main disadvantages is that the waste products are highly radioactive and are therefore very difficult to dispose of safely.
Two advantages of nuclear power are:

1 there are adequate supplies of uranium to last for a very long time;
2 nuclear power stations do not release gases such as sulphur dioxide and carbon dioxide which can harm the environment.

The accident at the nuclear reactor in Chernobyl, USSR, in 1986 was caused by the control rods being removed too far from the reactor core, so that the nuclear fission reaction produced large amounts of heat which could not be removed quickly enough from the reactor. The heat caused an explosion which exposed the top of the reactor core to the atmosphere, and ejecting large amounts of radioactive debris. Some of the radioactive substances were carried by strong winds across into Europe, where heavy rainfall caused contamination of areas of Scotland, the Lake District and North Wales.

NUCLEUS

NUCLEUS OF AN ATOM

The nucleus of the atom is positively charged, takes up a very small amount of space, yet contains nearly all the **mass** of the atom. It is the breakdown of radioactive nuclei that give rise to radioactivity (nuclear radiation).
Within the nucleus are found sub-atomic particles, namely **protons** and **neutrons**. The proton has a mass of 1u and a charge of +1; the neutron has a mass of 1u, but carries no charge.

NUCLEUS OF A CELL

The nucleus of a cell controls the activities of the cell. It contains the **chromosomes** and is surrounded by a nuclear membrane. All cells have a nucleus, except red blood cells.

NUTRIENT CYCLES

◀ Nitrogen cycle, carbon cycle, water cycle ▶

NUTRITION

This describes the process of taking in food and digesting it to provide energy for all the processes carried out by the cells, such as growth.
◀ Balanced diet ▶

NYLON

Nylon is a **polymer** which is manufactured from the two **monomers** of hexamethylenediamine and adipyl chloride. It is easily formed by pouring liquid hexamethylenediamine onto a solution of adipyl chloride. Where the two liquids meet, a film of nylon is formed which can be pulled out with tweezers as a continuous thread (Fig N.7).

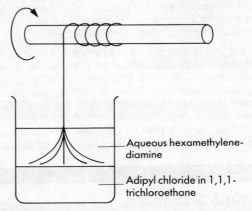

Fig N.7 The 'nylon rope trick'

Nylon has a structure similar to some proteins; it is strong, waterproof, wear-resistant and does not rot. It is a **thermosoftening plastic** so can easily be made into threads or shaped into solid objects. It can be used as fibres for clothes, shirts, carpets, ropes, and wear-resistant objects such as machine parts.

OESTROGEN

A female **hormone** released by the **ovary** which causes the development of secondary sexual characteristics, development of the breasts, fat deposits around the hips, and growth of hair under the arms and around the pubic area. Oestrogen also increases the thickness of the uterus wall ready for the implantation of the **zygote**.

OHM

The ohm is the **resistance** of a conductor in which the current is 1 **ampere** when a **potential difference** of 1 volt is applied across it. For example, one ohm of resistance is given by a resistor if a voltage of one volt is required to push a current of one amp through the resistor. So a higher resistance means more voltage is required. Resistance is measured in units called ohms, symbol Ω.

To calculate the resistance you need to know the voltage and current:

resistance = $\dfrac{\text{voltage}}{\text{current}}$

Alternatively, $V = IR$

or $R = \dfrac{V}{I}$

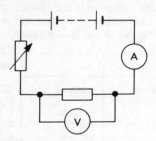

Fig O.1 A circuit to investigate Ohm's law

OHM'S LAW

The current through a metallic conductor is directly proportional to the **potential difference** across its ends, if the temperature and other conditions are constant.

You may have carried out a practical investigation to compare the relationship between the voltage and the amount of current flowing in a circuit, using a circuit similar to the one in Figure O.1. The graph in Figure O.2 shows how the current flowing (I) is directly proportional to the voltage (V), at constant temperature.

Figure O.3 is a useful way of remembering how to use this formula.

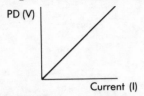

Fig O.2 Ohms' Law: a graph showing the relationship between current and voltage, using a resistor in the circuit

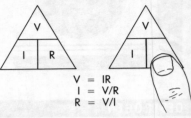

$V = IR$
$I = V/R$
$R = V/I$

Fig O.3 The ohm: remembering Ohms' Law

OMNIVORE

These are animals which obtain their energy by feeding on a mixed diet of plants and animals. They feed at more than one level in the food chain; for example, robins feed on seeds and berries in the winter, and in summer they feed on caterpillars and other insects. Humans obtain energy by eating both meat and vegetables.

OPTIC NERVE

◀ Eye ▶

ORGANIC COMPOUNDS

Organic compounds are divided into **homologous** series. The chart below gives the names of the first few members in each series together with their melting points and boiling points.

Series	Name	Formula	M.p.°C	B.p.°C
Alkanes	methane	CH_4	−183	−162
	ethane	C_2H_6	−172	− 89
	propane	C_3H_8	−187	− 42
	butane	C_4H_{10}	−135	− 0.5
	pentane	C_5H_{12}	−130	36
Alkenes	ethene	C_2H_4	−169	−102
	propene	C_3H_6	−185	− 48
	butene	C_4H_8	−135	− 7
Alcohols	methanol	CH_3OH	− 97	65
	ethanol	C_2H_5OH	−114	78
	propanol	C_3H_7OH	−126	97
Acids	methanoic acid	HCO_2H	9	101
	ethanoic acid	CH_3CO_2H	17	118

Gases

For a substance to be a gas, its melting point and boiling point must be below room temperature, say 20°C. Thus some gases in the above list are methane, ethene and butene.

Liquids

For a substance to be a liquid its melting point must be below room temperature and its boiling point above room temperature. Among the liquids in the above list are pentane, methanol and methanoic acid.

Solids

For a substance to be a solid its melting point and boiling point must be above room temperature. There are no solids in the above list, although if room temperature drops below 17°C, ethanoic acid becomes a solid.

ORGANISM

The word organism means any living animal or plant, including **bacteria** and **viruses**.

ORGAN SYSTEMS

In your body cells are grouped together to make tissues which in turn make organs which form organ systems. There are seven main organ systems in your body (Fig O.4):

1. The **circulatory system**, which carries oxygen, glucose and amino acids to every cell, and carries waste products such as urea and carbon dioxide away from the cells.
2. The **respiratory system**, which takes in oxygen and removes carbon dioxide.
3. The **digestive system**, which breaks down and absorbs the food taken into your body.
4. The **excretory system**, which removes unwanted, harmful waste produced by your body, such as urea produced by the liver and removed by the kidneys.
5. The **skeletal system**, which protects and supports your organs and muscles, and enables your muscles to move your body.
6. The **nervous system**, which controls all the organs in your body and enables your body to respond to the information received by its sensory cells.
7. The **reproductive system**, which enables you to make eggs or sperm so that you can pass on genetic information to create the next generation.

ORGAN SYSTEMS

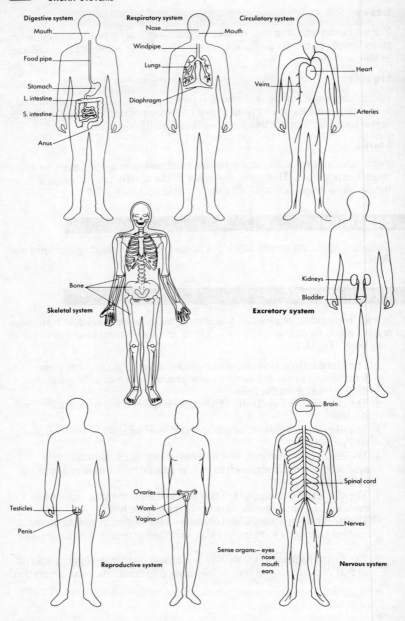

Fig 0.4 Organ systems

OR-GATE

◄ Logic gates ►

OSCILLATIONS

If you hang an object on a piece of string and let it swing backwards and forwards, then you are allowing it to *oscillate*. Eventually the oscillations slow down and the object comes to a stop. When you start it moving again the **amplitude** of the oscillations becomes larger and then smaller. One complete oscillation is from A to B to C and back to A, as shown in Figure O.5.

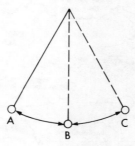

Fig O.5 Oscillations

The time taken for one complete oscillation is known as the *period* of the oscillation. A child on a swing making 10 complete swings or oscillations in 60 seconds has a period of oscillation of 6 seconds.

The number of oscillations made each second is the **frequency** of the oscillation. Frequencies are measured in cycles per second or **hertz** (Hz). If the child makes 10 complete swings or 10 cycles in 60 seconds then they make 0.16 cycles in 1 second. The frequency of the oscillation is 0.16 cycles per second or 0.16 hertz.

Oscillations and the body

Oscillations can have a variety of effects on body function:

- The oscillations produced by a ship which is rolling from side to side at sea can cause people to feel seasick.
- A drum beat which produces very low frequency oscillations can make people feel giddy and cause blurred vision.
- Sometimes people who work in factories are affected by the oscillations of the machinery which they operate.
- Musical instruments, such as a piano, produce sounds due to the oscillations of the piano strings.

OSMOREGULATION

Osmoregulation is the term we use to describe the process of maintaining the correct fluid balance in our bodies. On average there is about 58% water in an adult person so it is essential that the amount of water remains constant. If too much water is drunk the body fluids become dilute; if too much water is lost then the body fluids become too concentrated. Whichever of these situations arises, the body cells would cease to function properly (Fig O.6). The organ which controls the amount of water leaving the body is the **kidney**, which works with the **hypothalamus** and anti-diuretic hormone (ADH) to achieve **homeostasis**.

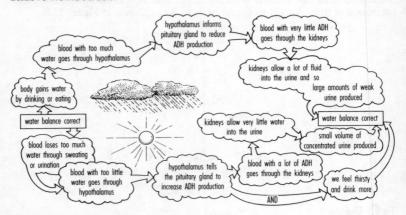

Fig O.6 Osmoregulation: how fluid level is controlled in mammals

OSMOSIS

The difference in size of **particles** can produce some interesting effects. For example, if a concentrated sugar solution is placed in a bag made of **visking tubing** (a material like cellophane), and the bag is placed in a beaker of dilute sugar solution, then after a few hours time the bag will increase in volume.

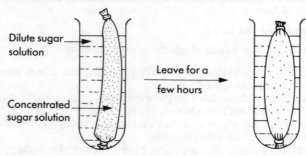

Fig O.7 Osmosis

This is because the visking tubing is acting as a sort of particle sieve. The tubing surface is covered with tiny holes or *pores*, each just big enough to allow small water particles to pass through, but **not** the large sugar particles (Fig O.8).

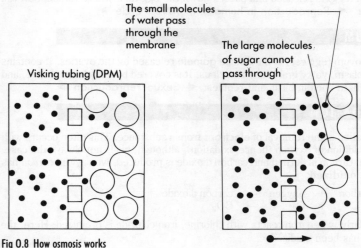

Fig O.8 How osmosis works

Materials such as visking tubing are called **differentially permeable membranes** (DPM). If the process above were allowed to continue, water would pass through the tubing walls until the concentration of the solutions inside and outside were the same. The movement of water from a weak to a strong solution through a DPM is called osmosis. This is the way in which water is passed from cell to cell in living things.

OSSICLES

The ossicles in mammals are three small bones of the middle ear which transmit vibrations from the ear drum to the fluid inside the *cochlea*. In birds, reptiles and amphibians there is usually only one ossicle. ◀ Ear ▶

OVARY

In animals the ovary is the female sex organ where the ova (eggs) are produced. **Oestrogen**, a female **hormone** is also released by the ovary.
◀ Sexual reproduction ▶.

OVERHEAD CABLES

◀ Transmission of electricity ▶

OVULATION

Ovulation is the release of an egg (ovum) from the **ovaries**. Every 28 days an ovum (egg) is released and passes down the egg tube where **fertilization** can occur. ◄ Reproduction in humans, menstrual cycle ►

OVUM

The ovum (egg cell) is the female **gamete** released by the ovaries. It contains cytoplasm, yolky grains and a nucleus. It is covered by a thick membrane, and by a hard shell in birds and reptiles. ◄ Sexual reproduction ►

OXIDATION

Oxidation is the removal of **electrons** from a substance. This often occurs with the addition of oxygen (hence oxidation), although this is not always the case. When carbon is burned and carbon dioxide is produced, we say the *carbon* has been **oxidized**:

carbon + oxygen → carbon dioxide
C + O_2 → CO_2

Similarly, when iron reacts with chlorine, iron chloride is produced. Here, the *iron* has been oxidized:

iron + chlorine → iron chloride
2Fe + $3Cl_2$ → $2FeCl_3$

The iron has been oxidized because electrons have been removed from its atoms to form Fe^{3+} ions:

Fe − $3e^-$ → Fe^{3+}

We need to understand oxidation when we talk about **corrosion** of metals and **combustion**, both of which are examples of oxidation.

Sometimes we refer to the *oxidation state* of a substance. This indicates how *much* it has been oxidized, i.e. how many electrons have been removed. For example, in the case of iron, the +3 state is more highly oxidized than the +2 state:

Atom/ion	Oxidation state
Fe	0
Fe^{2+}	+2
Fe^{3+}	+3

OXIDIZING AGENT

An oxidizing agent is a substance which is very good at 'pulling' electrons away from another substance or adding oxygen to a substance. Oxygen is a good oxidizing agent as are fluorine, chlorine and potassium manganate (VII). ◄ Oxidation ►

OXYGEN

Oxygen is a gas that makes up about one-fifth of the air around us. It is colourless and has no smell. Oxygen gas contains **molecules** of oxygen; its chemical formula is O_2.

Oxygen is a very important gas because it is needed to support living things (respiration) by reacting with food substances (sugars) to provide energy. The balance of oxygen in the air is maintained because when plants **photosynthesise** they produce oxygen as a by-product.

Oxygen is also the gas which allows things to burn; it reacts with the 'fuel' to produce an oxide and releases energy as heat. This is called *combustion*. Oxygen is very important in industry and is manufactured by separating it from the air. It is usually stored as *liquid oxygen* (by cooling it sufficiently to make it liquid). Rockets have to carry their own supply of oxygen in order to burn the fuel (hydrogen) which they also carry.

OXYHAEMOGLOBIN

◀ Breathing ▶

OZONE LAYER

The ozone layer surrounds the **Earth** in the part of the upper atmosphere known as the **stratosphere** (Fig O.9). The stratosphere is vital to life since it shields us from the most harmful types of **ultra violet (UV) radiation** from the sun. If more UV radiation penetrates, then the number of cases of skin cancer will increase. There is also strong evidence that increased exposure to UV would harm crops, affecting the world's food supply.

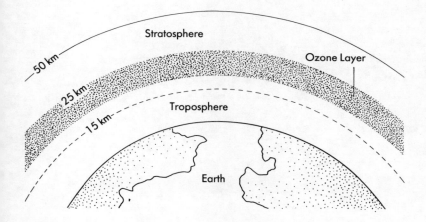

Fig O.9 The ozone layer

OZONE LAYER

Ozone is a form of oxygen which has three atoms in each molecule (O_3) compared with the usual form we breathe which has two atoms in each molecule (O_2).

In simple terms, ozone protects us by a series of chemical reactions. Some types of UV radiation will split the oxygen molecule (O_2) to form oxygen atoms. These are very reactive and can combine with oxygen molecules to form ozone (O_3):

$$O_2 \xrightarrow{\text{UV radiation}} O\cdot + O\cdot$$

$$O\cdot + O_2 \rightarrow O_3$$

The ozone produced then absorbs other types of UV radiation, which converts them back to oxygen molecules and atoms:

$$O_3 \rightarrow O_2 + O\cdot$$

Although these reactions protect us from the harmful UV radiation, their balance is being upset by the release of chemicals such as **chlorofluorocarbons** (CFCs) from **aerosols** etc., and oxides of nitrogen (from car exhausts), which also react with the ozone making fewer ozone molecules available at any time to absorb the UV radiation.

PANCREAS

◀ Digestion ▶

PARALLEL CIRCUIT

In a parallel circuit the components are connected to the same **electromotive force** (e.m.f.) so that the same **potential difference** is applied to each part.

Figure P.1 shows two lamps connected in a parallel circuit. Each lamp glows brightly as it has the total voltage across it. The lamps take twice as much current, so the reading of 0.2 amps on ammeter A_2 and A_3 is the same, but the total current, measured by A_1 and A_4, is 0.4 amps.

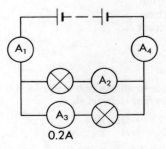

Fig P.1 Parallel circuit

The advantage of a parallel circuit is that the lamps glow brightly and if one lamp is faulty the others stay alight. This is especially useful in wiring decorative lights, for example, on a christmas tree, or along a street. Disadvantages of parallel circuits are that they can be complex to wire up correctly, and use up more wire which can be costly.

PARASITE

A parasite is an organism which lives in, or on, another organism and obtains food from it. Parasites do not usually kill their host but they may have a harmful effect on their host. For example, a flea living on the blood of a cat or dog, a protozoa living inside your gut.

PARTICLE

The word particle usually refers to the 'particles which make up all matter' (i.e. **solid, liquid** or **gas**). These particles include **atoms, molecules** and **ions**. They are so small that they cannot be seen even through the most powerful microscope. The particles which make up all matter are also moving.
◀ Kinetic theory ▶

PEPSIN

◀ Digestion ▶

PERIODIC TABLE

The periodic table is a complete list of all the **elements** and hence the **atoms** that exist. However, the list is arranged in a definite grid or pattern of rows and columns (Fig P.2). Each *row* is called a *period* and each *column* is called a *group*. The simplest atom (the lightest) appears at the top left and the most complex (the heaviest) appears at the bottom right. The atoms are arranged in increasing order of mass.

Atoms vary in complexity depending on how many **protons, neutrons** and **electrons** they have. Protons and neutrons are found in the nucleus; electrons are found orbiting the nucleus in 'shells'. Each shell can hold a set amount of electrons:

- 1st shell – 2 electrons
- 2nd shell – 8 electrons
- 3rd shell – 8 electrons etc.

▶ PERIODS

Each period corresponds to an electron shell. As you move across the *first period* you are filling the *first* electron shell:

atom:	H	He
atomic number:	1	2
electron configuration:	1	2

As you move across the *second period* you are filling the *second* electron shell:

atom:	Li	Be	B	C	N	O	F	Ne
atomic number:	3	4	5	6	7	8	9	10
electron configuration:	2,1	2,2	2,3	2,4	2,5	2,6	2,7	2,8

In the *third period* you are filling the *third* electron shell:

atom:	Na	Mg	Al	Si	P	S	Cl	Ar
atomic number:	11	12	13	14	15	16	17	18
electron configuration:	2,8,1	2,8,2	2,8,3	2,8,4	2,8,5	2,8,6	2,8,7	2,8,8

I	II											III	IV	V	VI	VII	O
1 H Hydrogen 1																	4 He Helium 2
7 Li Lithium 3	9 Be Beryllium 4											11 B Boron 5	12 C Carbon 6	14 N Nitrogen 7	16 O Oxygen 8	19 F Fluorine 9	20 Ne Neon 10
23 Na Sodium 11	24 Mg Magnesium 12											27 Al Aluminium 13	28 Si Silicon 14	31 P Phosphorus 15	32 S Sulphur 16	33.5 Cl Chlorine 17	40 Ar Argon 18
39 K Potassium 19	40 Ca Calcium 20	45 Sc Scandium 21	48 Ti Titanium 22	51 V Vanadium 23	52 Cr Chromium 24	55 Mn Manganese 25	56 Fe Iron 26	59 Co Cobalt 27	59 Ni Nickel 28	63.5 Cu Copper 29	65 Zn Zinc 30	70 Ga Gallium 31	73 Ge Germanium 32	75 As Arsenic 33	79 Se Selenium 34	80 Br Bromine 35	84 Kr Krypton 36
85 Rb Rubidium 37	88 Sr Strontium 38	89 Y Yttrium 39	91 Zr Zirconium 40	93 Nb Niobium 41	96 Mo Molybdenum 42	Tc Technetium 43	101 Ru Ruthenium 44	103 Rh Rhodium 45	106 Pd Palladium 46	108 Ag Silver 47	112 Cd Cadmium 48	115 In Indium 49	119 Sn Tin 50	122 Sb Antimony 51	128 Te Tellurium 52	127 I Iodine 53	131 Xe Xenon 54
133 Cs Caesium 55	137 Ba Barium 56	139 La Lanthanum 57 *	178 Hf Halfnium 72	181 Ta Tantalum 73	184 W Tungsten 74	186 Re Rhenium 75	190 Os Osmium 76	192 Ir Iridium 77	195 Pt Platinum 78	197 Au Gold 79	201 Hg Mercury 80	204 Tl Thallium 81	207 Pb Lead 82	209 Bi Bismuth 83	Po Polonium 84	At Astatine 85	Rn Radon 86
Fr Francium 87	236 Ra Radium 88	227 Ac Actinium 89 †															

140 Ce Cerium 58	141 Pr Praseodymium 59	144 Nd Neodymium 60	Pm Promethium 61	150 Sm Samarium 62	152 Eu Europium 63	157 Gd Gadolinium 64	159 Tb Terbium 65	162 Dy Dysprosium 66	165 Ho Holmium 67	167 Er Erbium 68	169 Tm Thulium 69	173 Yb Ytterbium 70	175 Lu Lutetium 71
232 Th Thorium 90	Pa Protactinium 91	238 U Uranium 92	Np Neptunium 93	Pu Plutonium 94	Am Americium 95	Cm Curium 96	Bk Berkelium 97	Cf Californium 98	Es Einsteinium 99	Fm Fermium 100	Md Mendelevium 101	No Nobelium 102	Lr Lawrencium 103

*** 58–71 Lanthanum series**
† 90–103 Actinium series

Fig P.2 Periodic table

PERIODIC TABLE

As you go *across* the table the atoms get bigger, since there are more electrons which take up most of the space. The atoms also get *heavier*, since there are more protons and neutrons (Fig P.3).

Metal/non-metal	Na m	Mg m	Al m	Si n/m	P n/m	S n/m	Cl n/m	Ar n/m
Outer shell electrons	1	2	3	4	5	6	7	8
Valency	1	2	3	4	3	2	1	0
Oxidation no.	+1	+2	+3	+4	−3	−2	−1	0
Melting point/°C	98	650	660	1410	44	113	−100	−189
Boiling point/°C	880	1100	2470	2355	280	444	−35	−186
Oxide nature	basic	basic	amphoteric	acidic	acidic	acidic	acidic	−
Formula of oxide	Na_2O	MgO	Al_2O_3	SiO_2	P_2O_3	SO_2	Cl_2O	−
Formula of chloride	$NaCl$	$MgCl_2$	$AlCl_3$	$SiCl_4$	PCl_3	S_2Cl_2	Cl_2	−

Fig P.3 Periodic table: trends across the 3rd period

In the same way, as you go down the table atoms get bigger and heavier. Notice also that as you go *down* a group, the atoms have the *same number of electrons in their outer shell*. For example, in Group 1:

atom	electron configuration
H	1
Li	2, 1
Na	2, 8, 1
K	2, 8, 8, 1

This is important, since it is how electrons are arranged in their shells that determines how atoms behave in chemical reactions.

Trends down a group; families of elements

The elements in each *group* of the periodic table behave similarly in chemical reactions. (*Remember*: they have the *same number of electrons in their outer shells*.) However, they do differ by degrees in their reactivity as you travel 'down the group'. Elements within the same group are referred to as 'families':

Group 1 – The alkali metals   (a group of metals)
Group 7 – The halogens   (a group of non-metals)
Group 0 – The inert gases   (a special group)

When atoms react with each other they often do so to form ions. In general, metals form *positive ions*; non-metals form *negative ions*. The *size* of the charge on the ion can be predicted by its position in the periodic table. For example:

- Group 1 elements form ions with *one* positive charge.
- Group 2 elements form ions with *two* positive charges.
- Group 3 elements form ions with *three* positive charges.
- Group 7 elements form ions with *one* negative charge.
- Group 6 elements form ions with *two* negative charges.

Since the periodic table shows many other trends, the position of atoms within the table can be used to predict their properties (Fig P.4).

◀ Atomic structure ▶

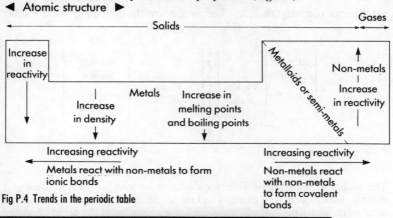

Fig P.4 Trends in the periodic table

PERISTALSIS

Wave-like movements of the intestines which propel the contents (i.e. the food) through the gastrointestinal tract.

PESTICIDES

Gardeners and farmers often use chemical pesticides to control insects which are damaging crops and other plants. Herbicides are also used to kill unwanted plants which otherwise affect the yield of crops. Although these chemicals may be used only in very small quantities, the concentration of chemical builds up each stage of the food chain, and accumulates in the top carnivore, for example, a bird of prey such as an owl or hawk. Because the chemicals affect each organism in the food chain, the top carnivore will receive the highest concentration, and be affected the most. It may be killed or have very low rates of reproduction.

The advantage of using chemical pesticides is that they are very effective and fast working.

PHOTOSYNTHESIS

Photosynthesis is the process in green plants which converts carbon dioxide and water into carbohydrates and oxygen. An equation for this process is:

$$\text{carbon dioxide} + \text{water} \xrightarrow[\text{chlorophyll}]{\text{sunlight}} \text{carbohydrates} + \text{oxygen}$$

$$6CO_2 + 6H_2O \longrightarrow C_6H_{12}O_6 + 6O_2$$

Energy from the Sun is absorbed by *chlorophyll* (the green pigment in plant leaves) and used to make sugars which are stored as starch.

pH SCALE

pH is a scale of acidity/alkalinity. The numbers on the scale range from 1 to 14:

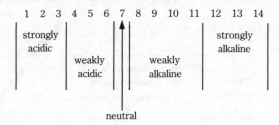

The acidity of a particular substance or solution can be measured using a suitable indicator which changes colour at different pH values, or more accurately, with a pH meter. These are some examples of the pH of everyday materials: lemon juice has a pH of 2; pure water is pH 7; ammonia solution has a pH of 11.

pH is really a measure of the hydrogen ion (H^+) concentration in the solution. It is the H^+ ion which gives rise to acidity. High hydrogen ion concentrations (i.e. strongly acidic) have *low* pH numbers.

PITUITARY GLAND

This is an *endocrine gland* which secretes a number of different **hormones**, to control the action of other endocrine glands. It is known as the 'master' gland of the body and is the most important endocrine gland.

PLANET

A **planet** is a body which orbits the Sun, for example the Earth, Mercury, Jupiter. Planets do not emit light but they reflect light from the Sun, and therefore appear as 'stars' in the night sky. ◄ Solar system ►

PLASTICS

Plastics form a group of synthetic **polymers** which are particularly mouldable, especially at high temperatures. In plastics the polymer molecules are very long chains. Plastics are manufactured with the object of building up **compounds** with predicted properties; they can be tailor-made.

The properties depend on the degree of polymerization (length of the chain). Polymerization, to produce polymer chains of known length can be done in various ways, e.g. varying the amounts of *catalyst* used or adding a chemical called an *inhibitor* to stop the reaction at a certain time. Plastics are manufactured in a variety of forms, e.g. powder, granules or sticky liquids, and converted into a final product, often by heating and moulding.

POLLUTION

There are two main types of plastics, **thermosoftening plastics** and **thermosetting plastics**. Each type has specific uses (Fig P.5).

Thermosoftening plastics	Uses
Polyethene (Polythene)	bags, films for packaging toys, household goods, insulation for electrical wiring
Polypropylene	tableware, chair seats, toilet seats, heels for shoes, filaments for brushes
PVC	water pipes, drain pipes, packaging, gramophone records, coating fabrics, rainwear, floor tiles
Polystyrene	household containers, toys, expanded foam insulating material, packaging
Polyester	clothes, sheets, ropes, tents, sails, safety belts

Thermosetting plastics	Uses
Bakelite	electrical switch and plug covers, bottle and container tops, door handles, ash trays
Urea-fomaldehyde resins	adhesives, surface coatings of metal, laminating timbers
Melamine-formaldehyde resin	moulding, laminated sheet, table ware
Polyester resin	reinforced with glass fibres used in boat and car body production, crash helmets, varnish and paint
Polyurethanes	in foam form for sponges, cushions, buoyancy in boat hulls

Fig P.5 Uses of plastics

Advantages of plastics

- they are cheap and easy to mould into shapes;
- they do not corrode and can resist chemical attack;
- they are waterproof;
- they can be easily coloured;
- they are lightweight.

Disadvantages of plastics

- their manufacture uses raw materials derived from oil (a non-renewable resource);
- they are flammable and give off toxic fumes when they burn;
- they cannot easily be disposed of, waste plastics in land-fill sites will not decompose in the soil. ◄ Biodegradable ►

POLLUTION

Any harmful substances which enter the environment can be described as pollution, especially if they reach unusually high levels.

For example:

1. *Air pollution*: dust, smoke, soot, **sulphur dioxide**, carbon monoxide, lead oxide, nitrogen oxides. Mostly produced by burning fossil fuels, such as coal or petrol.
2. *Water pollution*: oil, detergent, sewage, fertilizers, industrial waste.
3. *Soil pollution*: dumping of rubbish, chemicals, cars, radioactive waste.
4. *Noise pollution*: aeroplanes, engines, industrial machinery, motorbikes, loud music.

POLYMERS

Polymers are substances which consist of **molecules** with very long chains. Each chain is made up of repeating monomer units. There can be between 1000 – 50,000 **monomers** in a chain (Fig P.6). A monomer is a small molecule which when joined with others forms the polymer molecule.

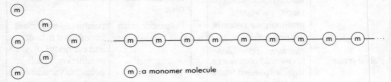

Fig P.6 A polymer molecule: polymers are made by joining together many small molecules called monomers

There are *natural polymers* found in living things and *synthetic polymers* (e.g. plastics):

- natural polymers: starch, proteins, wool, cotton, cellulose etc.
- synthetic polymers: **polythene**, bakelite, polytetrafluorethylene (PTFE or Teflon) polyvinylchloride (PVC) etc.

POLYTHENE

Polythene is a synthetic **polymer** manufactured by the process of *addition polymerization*. The raw material for the manufacture of polythene is ethene gas (ethene is the monomer). The polymerization process can be represented as:

$$n\ CH_2 = CH_2 \xrightarrow[\text{high temperature and pressure}]{\text{initiator}} \text{\textendash}[CH_2\text{\textendash}CH_2]\text{\textendash}\ n$$

Where n is a very large number.

Polythene can be made in a high density form or a low density form, each with different properties. High density polythene is used to make containers, pipes and kitchenware. Low density polythene is used for food wrapping, film and bags.

POPULATIONS

A population is the number of organisms of the same species in a particular area. For example, the number of snails in a garden, the number of oak trees in a woodland.

Figure P.7 shows a typical growth curve for a population. The new population starts with low numbers in the 'lag phase', and then shows a very rapid increase in number in the 'log' phase. When resources limit growth then a stabilisation phase is reached and the population remains fairly constant, until there is a change in one of the limiting factors, such as availability of food, space or disease.

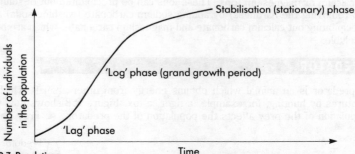

Fig P.7 Populations

POSITIVE FEEDBACK

◀ Feedback systems ▶

POTENTIAL ENERGY

Potential energy is a form of energy which is stored; for example, the energy stored in a wound-up spring, the energy stored in an object which is raised up.

POWER

Power is the rate of doing **work**, and is measured in **joules** per second (J/s) or watts (W). The formula for power is

$$\frac{\text{amount of work done}}{\text{time taken to do the work}}$$

For example, if a person does 50 joules of work in 10 seconds then their rate of working is $\frac{50}{10} = 5$ J/s or 5 W.

PRECIPITATION

Precipitation occurs when solid matter falls to the bottom of a solution; we say a *precipitate* has been formed. This may happen when two solutions are mixed together; some reactions between solutions of metal salts occur because a *solid* precipitate may be formed:

potassium iodide + lead nitrate → potassium nitrate + lead iodide

$2KI(aq) + Pb(NO_3)_2(aq) \rightarrow 2KNO_3(aq) + PbI_2(s)$

Notice that the ions have 'swapped partners'; such a change occurs because one possible combination, PbI_2, is insoluble in water.

Precipitation in action

Precipitation can be used to remove **hardness** from water due to the presence of calcium or magnesium **ions**. These ions can be precipitated out of solution (so removing the hardness) by adding sodium carbonate (washing soda) and precipitating out calcium carbonate and magnesium carbonate, which are both insoluble.

PREDATOR

A predator is an animal which obtains energy from other animals which it captures by hunting, for example, a tiger, a fox. Figure P.8 shows how the population of the prey affects the population of the predator. ◄ Prey ►

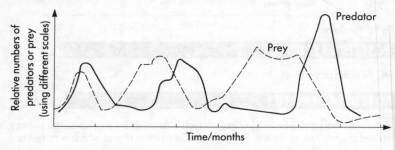

Fig P.8 Predator/Prey: the population of the prey affects the numbers of the predator

PREY

The prey is an animal which provides food for a **predator** which hunts and kills the prey. For example, a rabbit being killed by a fox. (Figure P.8).

PRIMARY COLOURS

Red, green and blue are the three primary colours which cannot be made by mixing any other colours. All the other colours can be made from two or three of these primary colours, as shown in Figure P.9.

When you mix two primary colours on a white screen, a new colour, called a *secondary* colour is produced. For example mixing red and green light produces yellow light. If the third primary colour, blue is now mixed with the yellow, white light is produced.

PROTEIN

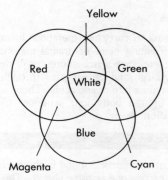

Fig P.9 Primary colours

PRIMARY CONSUMER

Herbivores are primary consumers. They obtain energy directly from green plants. For example, a cow or rabbit feeding on grass is a primary consumer.

PRODUCERS

Producers are green plants in a **food chain** which obtain their energy from the sunlight. They absorb light energy in the chlorophyll and use it to convert carbon dioxide and water into carbohydrates during the process of **photosynthesis**.

PROTEIN

A protein is a naturally-occuring **polymer** which is made of small monomer units called amino acids. There are about 20 different naturally-occuring **amino acids**, each containing a nitrogen atom. Different combinations of these amino acids can be combined in a polymer to make up different proteins (Fig P.10).

Plants manufacture proteins using sugars and starches (produced by **photosynthesis**), together with nitrates and water (absorbed through their roots) as raw materials. The nitrates provide the nitrogen atom. Proteins are used to build and repair cell walls.

Each acid contains $- NH_2$ group (amino)　　eg $H_2N -$ ▭ $- COOH$

and $- COOH$ group (acid)　　　　　　　or $H_2N -$ △ $- COOH$

The different shapes represent different amino acids

A typical protein

Fig P.10 Protein: proteins are polymers, the monomer units are amino acids

Proteins in action

Humans need proteins to carry out metabolic functions, and our source of protein comes from eating plant or animal material. Some foods such as cheese, eggs, meat etc., are a rich source of protein. Because the protein molecules are very large and cannot pass through the gut wall, they are broken down by enzymes into amino acids. These are absorbed and may be reassembled later by the body into proteins. (*Note*: enzymes are also proteins.) ◄ Digestion ►

PROTON

The proton is a subatomic particle found in the **nucleus** of an **atom**. It has a mass unit of 1u and carries one positive charge. The number of protons in an atom is given by the atom's **atomic number**. There are always the same number of **electrons** (each of which has a negative charge) as protons in an atom, so atoms are electrically neutral.

PULMONARY ARTERY/PULMONARY VEIN

The pulmonary artery carries deoxygenated blood from the right ventricle of the **heart** to the lungs.

The pulmonary vein carries oxygenated blood from the lungs to the left atrium of the heart.

PYRAMID OF BIOMASS

The amount of energy stored in a **food chain** can be found by weighing all the living organisms at each level of the food chain to find the total mass or biomass of organisms. The biomass decreases along the food chain, because less energy is available at each stage. A pyramid of biomass is shown in Figure P.11.

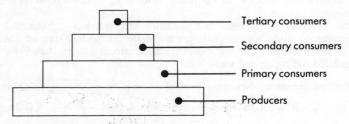

Fig P.11 Pyramid of biomass

PYRAMID OF NUMBERS

At each stage of the food chain there are usually many more organisms lower down the **food chain**. If you were to count all the living organisms in a food chain there may be thousands of **producers**, supplying energy for hundreds of **herbivores** which in turn supply energy for a few **carnivores**.

PYRAMID OF NUMBERS

At each stage of the food chain there is a decrease in the number of organisms as there is less energy available. A pyramid of numbers is shown in Figure P.12.

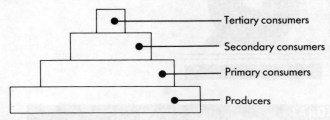

Fig P.12 Pyramid of numbers

Sometimes a pyramid of numbers can look like that in Figure P.13, where one organism, an oak tree, provides energy for many caterpillars, which provide energy for a few shrews, and in turn provide energy for one owl.

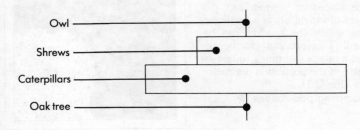

Fig P.13 A Pyramid of numbers based on one oak tree

QUADRAT

A quadrat is an open, square frame, usually made of wood or wire, which can be used in fieldwork for marking out a measured area of ground to be studied. For example, a $¼m^2$ quadrat can be placed at random and the numbers of individuals of a particular species inside the quadrat can be counted. This process can be repeated a number of times to find an estimate of the population per metre square (Fig Q.1).

◀ Sampling populations ▶

Quadrat

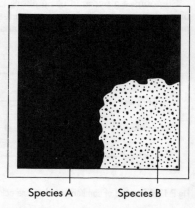

Species A   Species B

Fig Q.1 Using a quadrat to estimate percentage cover

RADAR

Radar stands for Radio Detection And Ranging. **Electromagnetic waves** are reflected from distant objects, and the reflections are recorded on a screen. Radar is widely used in air traffic control, shipping and weather forecasting.
◄ Satellites ►

RADIOACTIVITY

Radioactivity is caused by the spontaneous breakdown of some nuclei in atoms which are unstable. Those **isotopes** which are unstable (called radioisotopes) emit energy in the form of heat and radiation to become more stable. In the process the atom decays into an atom of a different element. The *rate of decay* is measured by the term **half-life**, the time taken for half the atoms to disintegrate.

For example, carbon-14 is a radioisotope whose nucleus contains 6 protons and 8 neutrons (the basic atom of carbon has only 6 neutrons and is not unstable). When the carbon-14 nucleus breaks down it emits energy and decays into the isotope nitrogen-14. Some radioisotopes (e.g. carbon-14) are naturally-occuring; others such as plutonium-239 can be manufactured. Naturally-occuring radioisotopes are usually found in the heavier elements, or are the isotopes of lighter elements which have more neutrons present in their nuclei.

 TYPES OF RADIOACTIVITY

When the radioactive nucleus breaks down it can emit up to three types of radioactivity (Fig R.1):
- **alpha** (α) particles – fast-moving helium nuclei;
- **beta** (β) particles – fast-moving electrons;
- **gamma** radiation (γ) – a form of electromagnetic radiation with very short wavelength.

Natural radiation

We are constantly exposed to radioactivity from natural sources referred to as **background radiation** arising from cosmic rays penetrating the atmosphere, from soil, rocks etc.

RADIOACTIVITY

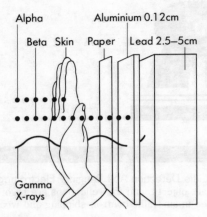

Fig R.1 The three types of radioactivity

Detecting radioactivity

Radioactivity was first discovered because of its ability to 'fog' photographic plates, in the same way that light affects photographic film. This is the way **X-ray** photographs are taken today. Radioactivity also ionizes gases through which it passes; this is the basis of modern detection methods, such as the **Geiger-Muller** tube. The tube is filled with gas, mainly argon. As radiation passes into the tube it ionises some of the gas atoms, causing a tiny electric current which can be measured. Ion production is directly related to radiation levels.

BIOLOGICAL EFFECTS

The biological effects of radiation on living tissue depends several factors:

1 the strength of the radiation;
2 the length of exposure;
3 how much of the tissue (how many cells) is exposed.

There may be no serious effect if only a few cells are damaged, although a plant or animal may die if enough cells are killed. If the radiation dose is high enough a **cancer** may develop in animals. There may also be genetic **mutations** causing future offspring to be different from their parents.

USES

Radioisotopes can be used in a variety of ways in industry, in medicine, in food production, or in radiocarbon dating:

1 By measuring the amount of radiation that passes through a material, a

RADIOACTIVITY

manufacturer can check the thickness of metal, the amount of toothpaste in a tube, soap powder in a packet etc. (Fig R.2).

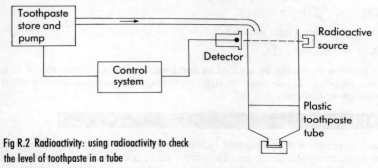

Fig R.2 Radioactivity: using radioactivity to check the level of toothpaste in a tube

2. All living things contain a large amount of carbon, most of which is carbon-12; a small proportion will be carbon-14. The proportion of carbon-12 to carbon-14 is the same for all living things, but when an organism dies, the amount of carbon-14 decreases (half-life 5570 years). By measuring the amount of radioactive carbon left one can date the item by reference to the half-life curve. This technique of carbon-14 dating was recently used to date the Turin Shroud.
3. Small amounts of radioisotopes can be introduced into underground water systems. Geiger counters can then be used to detect the position of leaks.
4. Food can be preserved by directing radiation (usually gamma rays) onto fresh food. This process can:

   - destroy bacteria and prevent the growth of moulds;
   - sterilize the contents of sealed packets;
   - reduce the sprouting of vegetables and prolong the ripening of fruits.

   This irradiation does *not* make the food radioactive, but cannot be used for all food since it can change the taste.
5. Controlling pests: large numbers of male insects are reared in the laboratory and sterilized by exposing them to a controlled dose of gamma rays. They are then released into the wild where they mate. However, the females with which they mate do not produce any young, so the insect population is quickly reduced.

▶ RADIOACTIVE DECAY

Radioactive decay is the spontaneous breakdown of an unstable **nucleus**, which can then emit either **alpha**, **beta**, or **gamma** radiation.

Alpha (α) particles can be represented as $^{4}_{2}\text{He}$.

Beta (β) particles can be represented as $^{0}_{-1}\text{e}^{-}$.

Gamma (γ) rays are electromagnetic radiation of short wavelength.

When radioactive atoms *decay* and emit particles they change into other atoms. For example, uranium-238 loses an alpha particle:

$$^{238}_{92}U \rightarrow\ ^{234}_{90}Th + ^{4}_{2}He$$

Carbon-14 loses a beta particle:

$$^{14}_{6}C \rightarrow\ ^{14}_{7}N + ^{0}_{-1}e^{-}$$

Radioactive atoms can be created by bombarding non-radioactive atoms with other particles. These could be alpha particles, beta particles or more often fast-moving neutrons.

RADIOWAVES

Radiowaves are **electromagnetic** radiation and have the longest wavelength in the electromagnetic spectrum. They are used to transfer information such as sound and TV pictures over long distances. Typical wavelength values are shown below:

Waves	Long	Medium	Short	vhf	uhf	micro
Wavelength	1500 m	300 m	30 m	3 m	30 cm	3 cm
Frequency	200 kHz	1 MHz	10 MHz	10^2 MHz	10^3 MHz	10^4 MHz
Use	◄─────────── Radio ───────────►			◄── T.V. ──►		Radar

Long, medium and short waves are used for local broadcasting as they can be bent around obstacles such as hills. Very high frequency and ultra-high frequency waves have shorter wavelengths and need a clear path from the transmitter to the receiver. ◄ Microwaves ►

RATE OF REACTION

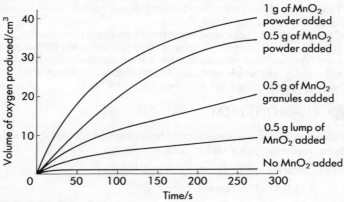

Fig R.3 Rate of reaction

REACTIVITY SERIES

The *speed* at which a reaction takes place can vary and depends on a number of factors (Fig R.3):

1. the surface area of any solid reactants;
2. the concentration of reactants (including pressure in the case of gases);
3. the temperature;
4. the presence of a catalyst.

In order for a chemical reaction to take place the particles of the reactants must collide (bump into each other). The more often the particles collide the more likely they are to react and so the faster the reaction.

INCREASING REACTION RATES

The speed of a reaction can be increased in a number of ways. Consider the reaction between calcium carbonate and hydrochloric acid to produce calcium chloride, water and carbon dioxide:

$$CaCO_3 + 2HCl \rightarrow CaCl_2 + H_2O + CO_2$$

Increasing the surface area of solid reactants

A greater surface area provides more opportunities for particles to collide. Powdered calcium carbonate reacts much faster than lumps of calcium carbonate (marble chips) because its surface area is much greater and more calcium carbonate is in contact with the acid. Stirring the powder in the acid will further increase the speed of reaction for the same reason.

Increasing the concentration of the reactants

This increases the number of particles present and so increases the chance of any collision. If the concentration of the hydrochloric acid is increased there is more chance of the particles interacting and the reaction will proceed at a faster rate.

Increasing the temperature of the reactants

Increased temperature provides the particles with more kinetic energy so they will move faster, increase the number of collisions per second and hence increase the rate of the reaction.

◀ Catalysts ▶

REACTIVITY SERIES

Metals *react* to form positive ions; some metals are *more reactive* than others, since less energy is required to remove electrons from their atoms i.e. they form **ions** more easily). Metals can be placed in order of their reactivity, based on their ability to form ions; this order is their reactivity series and tends to be the same, no matter how, or with what, the metals are reacting:

REACTIVITY SERIES

```
potassium                     greatest tendency to form ions
sodium
lithium
calcium
magnesium
aluminium

zinc                          increasing tendency to form ions
iron
tin
lead

copper
silver
gold                          least tendency to form ions
```
Reactivity series

The reactivity series helps to explain:

- why some, but not all, metals corrode;
- how we can prevent corrosion;
- why some metals and not others react with dilute acids;
- why some metals can be extracted from their ores by reduction with carbon, yet others can only be extracted by electrolysis;
- why metals were discovered in the order that they were.

REACTIONS OF METALS

Metals and oxygen

When heated in air, metals form *oxides*, which are bases. The more reactive metals (those near the top of the reactivity series) will burn in air. Gold is the only metal which is not oxidized by heating in air. For example:

$$2Mg(s) + O_2(g) \rightarrow 2MgO(s)$$

Metals and water

The more reactive metals react with water to produce hydrogen. Potassium, sodium, lithium and calcium react with cold water, producing hydroxides:

$$2Na(s) + 2H_2O(l) \rightarrow 2NaOH(aq) + H_2(g)$$

Magnesium, zinc and iron will react with steam producing oxides:

$$Zn(s) + H_2O(g) \rightarrow ZnO(s) + H_2(g)$$

Metals and acids

All the metals in the reactivity series above copper will react with dilute acids.

REACTIVITY SERIES

The lower in the reactivity series the metal the slower the reaction. For example:

$$Mg(s) + 2HCl(aq) \rightarrow MgCl_2(aq) + H_2(g)$$

Note: dilute nitric acid does not react in this way.

Metals and metal salt solutions

If a metal is placed in a solution of the salt of another metal a reaction may occur. Whether or not this will happen depends on the metal's position in the reactivity series. A metal *high* in the series will *displace* a metal *lower* in the reactivity series from solution. The metal high in the reactivity series also has the greater tendency to form ions:

$$Mg(s) + CuSO_4(aq) \rightarrow MgSO_4(aq) + Cu(s)$$

Magnesium is above copper in the reactivity series, so has a greater tendency to form ions; electrons are transferred from the magnesium atom to the copper ion. This can be shown as an *ionic equation*, since it does not matter what negative ion (e.g. nitrate or chloride) is present:

$$Mg(s) + Cu^{2+}(aq) \rightarrow Mg^{2+}(aq) + Cu(s)$$

Metals and cells

The difference in the different metals' abilities to form ions can be very useful. If two different metals are placed in a solution containing ions and the metals are linked by a wire, then electrons will flow through the wire. This means that a *current* is flowing through the wire, and there is a voltage between the two metals. This arrangement is called a simple cell. ◀ Cell – electrical ▶

The voltage that is produced between the two metals generally depends on their related positions in the reactivity series. If the metals are far apart in the series (e.g. magnesium (copper) a large voltage is produced; if they are close together (e.g. iron/zinc) a small voltage is produced. Refer back to the chart on page 210.

Metals and corrosion

Corrosion of a metal is a chemical reaction, and will only occur if the metal is in contact with a solution containing ions. As a metal corrodes it loses electrons to form positive ions:

Metal atom   – electron(s) → metal ion
    M   – e$^-$ → M$^+$

Metals corrode at different rates, depending on their position in the reactivity series; e.g. magnesium will corrode more quickly than copper.

History of metals

The most abundant metal in the Earth's crust is aluminium, yet it was one of the last to be discovered. Why should this be so? The reason for this lies in the reactivity series.

Metal	Approximate date of first use
gold, silver, copper	5000 BC
tin	2,500 BC
iron	1,200 BC
zinc	BC/AD
aluminium	1825 AD

If we compare the dates of discovery with the reactivity series we can see that one is the reverse of the other. In other words, the first metals discovered were those with the *least* tendency to form ions. Gold, silver and small amounts of copper can be found in their *native state* i.e. as the metals themselves. Other metals only exist in the Earth's crust as compounds. i.e. the metals are present as metal ions, chemically bonded with other substances. The more reactive a metal (the higher in the reactivity series) the more stable it becomes as a compound, so that it is difficult to extract it from its ore.

RECEPTORS

Receptor organs contain sensitive cells which detect changes in stimuli, such as heat and light, and send impulses along a sensory nerve. For example, the retina of the eye contains receptor cells which are sensitive to light and send impulses via the optic nerve to the brain. ◄ Reflex arc ▶

RECESSIVE

The gene which is recessive has *no effect* on the appearance of an individual who has both the **dominant** and recessive gene. An individual who is **homozygous** for a characteristic has two recessive genes and this *will* affect the appearance of the individual. For example, brown eye colour is dominant over blue eye colour. A person who is **heterozygous** has the gene from brown eyes and the gene for blue eyes. They appear to have brown eyes. A homozygous person has *both* recessive genes for blue eyes and appears to have blue eyes.

RECTIFICATION

Electronic equipment such as radios often require direct current but are supplied by mains AC. It is necessary to smooth out (*rectify*) the AC supply to give a steady DC current.

A **diode** placed in a circuit removes the negative half-cycle of the AC input to give a one-way **potential difference** across the equipment being used. Figure R.4a) shows half-wave rectification. Four diodes in the circuit form a bridge rectifier, which rectifies or smooths both half-cycles of the AC input

(Fig R.4b)). The current flows in the direction of the solid arrows when X is positive and Y is negative. When X is negative and Y is positive the current flows in the direction of the broken arrows, as shown.

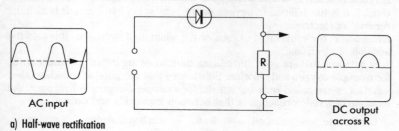

a) Half-wave rectification

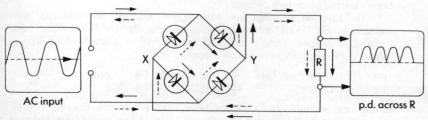

b) Full-wave rectification

Fig R.4 Rectification

RECYCLING

Many manufactured items can now be recycled. For example, glass bottles are crushed and melted down to make new glass containers; old cars are crushed and useful metal is extracted to make other consumer goods; newspapers are pulped and reprocessed to make 'new' paper.

RED BLOOD CELLS

Red blood cells contain *cytoplasm* surrounded by a cell membrane, but no nucleus. They are shaped like flattened discs and contain a pigment, **haemoglobin**, which makes the cell look red (Fig R.5). Haemoglobin combines with oxygen to form *oxyhaemoglobin*, to transport oxygen to the cells of the body from the **capillaries** surrounding the lungs. There are about 25 billion red blood cells in your body; about 5 million for every 1mm^3 of blood.

Fig R.5 Red blood cell (magnified × 200)

REDOX REACTIONS

A redox reaction is one in which both oxidation and reduction take place.
Oxidation is the addition of oxygen or the removal of hydrogen. It is also the removal of electrons.
Reduction is the removal of oxygen or the addition of hydrogen. It is also the addition of electrons.

Substances that are good at oxidising substances are called *oxidizing agents*, for example oxygen and chlorine. Substances that are good at reducing other substances are called *reducing agents*, for example carbon and hydrogen. An example of a redox reaction is that between lead oxide and carbon:

Lead oxide + carbon → lead + carbon dioxide
$2PbO$ + C → $2Pb$ + CO_2

The lead oxide has been reduced to lead by the carbon, while the carbon has been oxidized to carbon dioxide.

In the reaction between iron and copper sulphate, the iron has been oxidized to Fe^{2+} ions, whilst at the same time the Cu^{2+} ions have been reduced to copper atoms (Cu). This reaction takes place because iron is higher in the rectivity series than copper:

iron + copper sulphate → iron sulphate + copper
Fe + $CuSO_4$ → $FeSO_4$ + Cu

In the next example the iron has been oxidized to iron chloride. Three electrons have been removed from the iron atom. Chlorine has been reduced to the chloride ion, one electron has been added to each of the three chlorine atoms.

iron + chlorine → iron (III) chloride
$2Fe$ + $3Cl_2$ → $2FeCl_3$

Redox reactions in action

Oxidation cannot take place without **reduction** in a redox reaction since it involves the transfer of **electrons**.

The theory of **reduction** helps us to understand how metals are extracted from their ores. Carbon (as coke) is used as a reducing agent for some metals, for example iron and zinc.

The theory of **oxidation** helps us to understand the corrosion of metals. **Corrosion** involves the production of metal ions from their atoms (i.e. oxidation). For example, iron rusts to form iron oxide in the presence of oxygen and water:

$Fe - 3e^- \rightarrow Fe^{3+}$

REDUCING AGENT

A reducing agent is a substance which 'gives' electrons, adds hydrogen, or removes oxygen from another substance.

REFLECTION

Carbon is a good reducing agent used in the extraction of metals from ores; hydrogen is also a good reducing agent. Metals near the top of the **reactivity series** (those with the greatest tendency to lose electrons and form ions) are also good reducing agents.

REDUCTION

Reduction is the addition of **electrons** to a substance. This often occurs with the removal of oxygen or the addition of hydrogen. When carbon is heated with lead oxide it removes the oxygen from the lead oxide, which is *reduced* to lead:

lead oxide + carbon → lead + carbon dioxide
$2PbO$ + C → $2Pb$ + CO_2

Note: the lead ions Pb^{2+} in PbO have been reduced to lead atoms Pb, i.e. electrons have been added.

REFLECTION

When a wave hits a barrier the direction of the wave changes and it is reflected away from the barrier. If a plane wave hits the barrier at right angles it 'bounces back' along its original path, at the same wavelength, frequency and velocity. If the wave hits the barrier at an angle, it is reflected at the same angle away from the barrier, as shown in Figure R.6.

The angle of incidence equals the angle of reflection, and the wavelength, frequency and velocity stay the same.

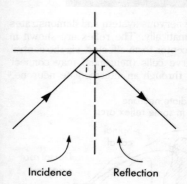

Fig R.6 Reflection of waves

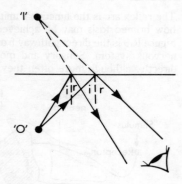

Fig R.7 Reflection of light

▶ REFLECTION OF LIGHT

You see the world around you because light rays are reflected from different objects into your eyes. When you look at yourself in the mirror you see an *image* of yourself reflected in the mirror. Your image appears as far behind the mirror as you are in front of the mirror. The image is described as a *virtual* or imaginary image. It is the same way up as you are, but the left and right sides are reversed. Figure R.7 shows how an image is formed by a plane mirror.

Figure R.8 shows a practical use of the reflection of light. An observer in an underground shelter uses a periscope to see above the ground.

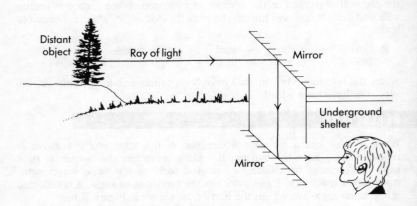

Fig R.8 Reflection of light: using a periscope

REFLEX ARC

The reflex arc is the functional unit of the nervous system and demonstrates how **homeostasis** may be achieved automatically. The reflex arc shown in Figure R.9 is the direct pathway from a receptor to an effector, via the central nervous system. Sensory and motor nerve cells (neurones) may connect directly, although more often they do so through an intermediate neurone.

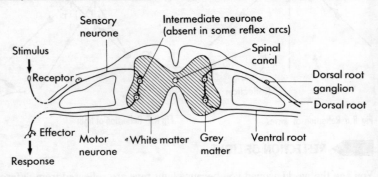

Fig R.9 Reflex arc

There are two main types of reflex pathway:

1 a simple reflex;
2 a conditioned reflex

A *simple* reflex consists of a reflex arc, and results in a very fast, automatic response to a stimulus. Simple reflexes are instinctive and usually increase an animal's chance of survival. Examples of simple reflexes are coughing, blinking and eye focusing, withdrawing a limb from a source of pain, and the knee jerk reflex (Fig R.10).

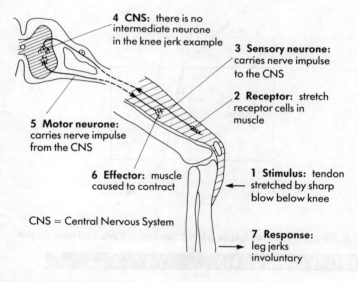

Fig R.10 A simple reflex: the knee jerk

A *conditioned* reflex involves learning and memory, and allows us to ride a bicycle or drive a car without too much conscious thought.

REFRACTION

Refraction is the bending of light rays as they pass from one medium to another. For example, waves travel at a certain speed in air, but when they pass into a different medium such as water, the speed slows down. The velocity of water waves decreases as they pass from deeper to shallower water, i.e. their speed slows down.

▶ REFRACTION OF LIGHT

When light waves pass into glass or water at an angle, their speed slows down and they change direction. It is this change of angle which is *refraction*.

If a ray of light enters a glass block at 90°, then it leaves the block in the same direction. It is when the light ray enters at an *angle* that its direction changes, both on entering and leaving the glass. Light refracts or bends towards the normal as it enters the glass which is more dense than the air. The light ray then refracts or bends away from the normal as it leaves the glass and passes into the air, a less dense medium (Fig R.11).

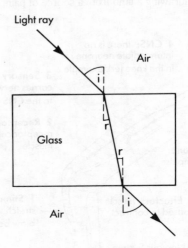

Fig R.11 Refraction of light: the light ray is bent or 'refracted' as it enters and leaves the glass

RELATIVE ATOMIC MASS

Different elements have different proportions of **isotopes** – atoms with different numbers of **neutrons** in their **nucleus** and which may therefore have a different **mass** to another atom of the same element.

The *relative atomic mass* of an element is based on the *average* mass of all the atoms in the element (taking the isotope carbon-12 as the standard) and will not be a whole number. It is the number usually quoted in a list of atomic masses, or given in the periodic table.

RELAY

A relay is an electrical device in which a small electric current controls a larger electric current by switching it on and off. ◀ Electromagnetic relay ▶

REPRODUCTION

Human male and female reproductive organs fit together during sexual intercourse so that the sperm (male gamete) are placed well inside the female's body. Sperm are made in the testes and pass out of the male's body in a fluid (semen) through the penis during ejaculation (Fig R.12a)). [They] have only a short distance to swim to enter the **uterus**. They move [up the] uterus and travel along the Fallopian tubes towards the **ovaries**. If [they meet an] egg in the Fallopian tube, **fertilization** may take place. The fer[tilized egg] travels to the uterus and embeds itself in the lining (endomet[rium] (Fig R.12b)). If fertilization does not take place, the endometrium breaks down and is lost as part of the monthly menstrual cycle. Gestation in humans is about 9 months; after birth human offspring require a great deal of parental care and attention.

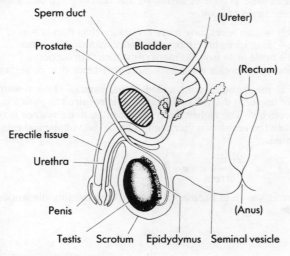

a) The human male reproductive system

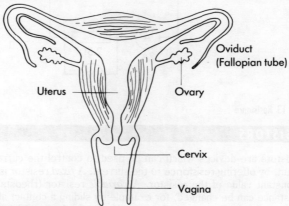

b) The human female reproductive system

Fig R.12 Reproduction

RESISTANCE

RESISTANCE

Resistance is a force which opposes the flow of an electric current so that energy is required to push the charged particles around the circuit. The circuit itself can resist the flow of particles if the wires in the circuit are very thin and very long, as in the case of a filament in an electric light bulb. Due to this resistance, energy is given out as heat and light. Many household appliances, such as electric heaters, hair driers, toasters, ovens and electric fires use a high resistance wire in their elements so that heat is given out. Four factors affect the resistance of a wire:

1. diameter – thin wires have more resistance than thicker wires;
2. length – long wires have more resistance than shorter wires;
3. what it's made of – iron has more resistance than copper;
4. temperature – hotter wires have more resistance than cooler wires.

Resistance is measured in units called ohms, symbol Ω. A resistor has a resistance of one ohm if a voltage of one volt is required to push a current of one amp through it. The higher the resistance, the more voltage is required.

To calculate the resistance you need to know the voltage and current, and use the formula:

$$\text{Resistance} = \frac{\text{Voltage}}{\text{Current}} \quad \text{or} \quad R = \frac{V}{I}$$

Use the circuit shown in Figure R.13 to see how the formula works.

◄ Ohm ►

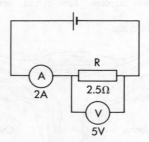

Fig R.13 Resistance

RESISTORS

Resistors are devices which can be used to control the current flowing in a circuit, by offering **resistance** to the current. A *fixed* resistor is where there is a constant value of the resistor. A *variable* resistor (rheostat) is where the resistance can be changed, for example by sliding a contact along a length of wire to vary the resistance. For example in a food mixer, the control knob can increase or decrease the speed of the mixer (Fig R.14). The knob is acting as a variable resistor and letting different amounts of current through. When the knob is turned to a high setting, the resistance is decreased and more current flows to the motor so the speed increases.

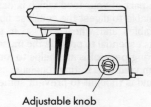

Fig R.14 Resistors: an everyday use of a variable resistor

RESPIRATION

Respiration is the release of energy from food, usually involving oxygen.
◀ Aerobic, anaerobic respiration, breathing ▶

RETINA

◀ Eye ▶

ROCK CYCLE

The rock cycle describes how rocks are continuously changing from one type into another over millions of years as shown in Figure R.15.
◀ Sedimentary, metamorphic, igneous rocks ▶

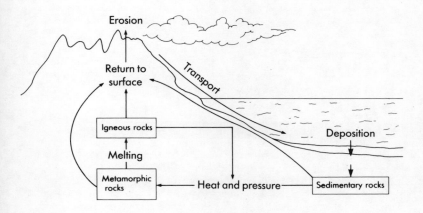

Fig R.15 Rock cycle; over millions of years rocks change from one type to another

ROUGHAGE

Roughage (fibre) is the part of your food which is essential for regular bowel movements. You are unable to digest the cellulose fibres of green plants such as apples, leafy vegetables and wheat and so it passes through the gut undigested. Roughage is essential to provide the muscles of the gut with some bulk to push against during **peristalsis** and helps to form soft faeces which can be eliminated easily from your gut. Eating roughage as part of a **balanced diet** prevents constipation and is thought to prevent some types of bowel cancer.

SACRIFICIAL PROTECTION

Metals can be protected from corrosion by Sacrificial protection. It is based on the fact that when two metals are in contact with each other the metal higher in the **reactivity series** will corrode.

If iron is coated with zinc, the zinc will corrode leaving the iron intact. This process of covering iron with a thin layer of zinc is called *galvanising*.

Ships and oil rig platforms can be protected from corrosion by welding large blocks of zinc to the hull, or to the legs of the oil rig (Fig S.1). The zinc will corrode in preference to the steel (iron) of the hull or legs. Replacing the zinc block when it has corroded is much easier and cheaper than replacing the steel.

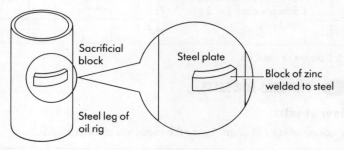

Fig S.1 Sacrificial protection

SALIVARY AMYLASE

◄ Amylase ►

SALTS

These are ionic substances formed in reactions between **acids** and **bases** or between acids and **metals**. When solutions of these salts are evaporated, crystals (giant ionic structures) are formed. Solutions of salts will conduct electricity, showing that they contain ions.

NAMING SALTS

The name of the salt depends on the ions present. Each salt contains a positive ion (cation), which is obtained from the metal, (e.g. Na^+, sodium; Mg^{2+}, magnesium) and a negative ion (anion), which is obtained from the acid (e.g. SO_4^{2-}, sulphate; CO_3^{2-}, carbonate). The name comes from a combination of both ions, for example, magnesium sulphate ($Mg^{2+}SO_4^{2-}$).

The charge on the metal ion depends on its group in the **periodic table**, although some metals (**transition metals**) can have ions with different charges (Fig S.2). The charge on the ion is indicated by roman numerals in the name. For example, iron (II) sulphate ($FeSO_4$) has the Fe^{2+} ion present; iron (III) sulphate ($Fe(SO_4)_3$) has the Fe^{3+} ion present.

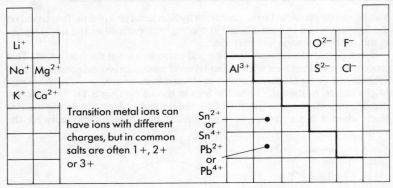

Fig S.2 Charges on metal ion

PATTERNS FOR SALTS

Colour of salts

The colour of salts is often due to the metal ion present (Fig S.3).

Ion	Colour	Ion	Colour
Na^+	colourless	Fe^{2+}	green
K^+	colourless	Fe^{3+}	red
Cu^{2+}	blue/green		

Fig S.3

Salts in action

Many salts are present in the sea and in the soil and provide plants with the elements they need. For example:

- nitrogen: nitrates (NO_3^-) in the soil provide the nitrogen which is essential for the plant to manufacture amino acids.

- phosphorus: phosphates (PO_4^{3-}) are essential for every energy transfer within the cell.
- potassium: (K^+) is needed for the activity of many enzymes.
- calcium: (Ca^{2+}) raw material for cell walls.
- magnesium: (Mg^{2+}) raw material for making chlorophyll.
- sulphur: sulphates (SO_4^{2-}) are raw material for making some amino acids.

Because these elements are needed by plants, their salts are manufactured as fertilizers. Examples are ammonium nitrate, ammonium phosphate and potassium chloride. Fertilizers containing these three salts are called N, P, K fertilizers because they provide nitrogen (N), phosphorus (P), and potassium (K).

Other salts can be used in a variety of ways. Calcium sulphate is used as wall plaster and plaster of Paris; silver chloride is used in photographic film emulsion; iron (II) sulphate is used in iron tablets to treat anaemia.

◄ Solubility ►

SAMPLING POPULATIONS

Sampling is a method of finding the size of a population which is too large to count. For example, if you were asked to find the numbers of dandelions growing in a large field, it would take too long to count each plant so sampling would be used instead. One technique used to estimate population size is described below:

1. Measure the whole area of study.
2. Use a **quadrat** of known size, e.g. $0.25m \times 0.25m = \frac{1}{4} m^2$
3. Place the quadrat at random.
4. Count the numbers of a particular species of animal, or assess the proportion of the quadrat which is covered by a particular plant.
5. Record your result.
6. Repeat stages 3, 4, and 5 until data has been collected from 10 quadrats.
7. Find the average numbers of the species, or average percentage cover of a plant per square metre.
8. Multiply by the total number of square metres to find the total population of the area.

SAPROPHYTES

Saprophytes are organisms which feed on dead or decaying animals and plants. Bacteria and fungi are saprophytes which decompose organisms and return the nutrients to the soil. ◄ Carbon cycle, nitrogen cycle ►

SATELLITES

The gravitational pull of the Earth provides the pull required to make a satellite follow a circular path around the Earth. For a satellite to orbit just above the atmosphere it needs to orbit at about 8,000 m/s. The Moon, a natural satellite of Earth, orbits much further away from Earth at a speed of about 1,000 m/s.

SATELLITES

Satellite pictures, like the one in Figure S.4 and on the TV weather news, are able to show the position and speed of the changing weather patterns. For example, a satellite could show the development of a cold front, or the position of a depression as it moved across the Atlantic Ocean, and how temperature varies vertically through the atmosphere. The information from satellites is used together with other information from more conventional methods of collecting data such as using hydrogen-filled weather balloons which float in the atmosphere, and computer weather forecasts. Meteorologists are then able to pass information to national and local news stations as well as to aeroplanes and shipping.

Fig S.4 Satellite picture; showing a frontal system approaching Britain

The type of satellite used in weather forecasting is a *geostationary* satellite whose speed is the same as that of Earth and is held in orbit by the Earth's gravitational field (Fig S.5). Some satellites are able to use both infra-red and visible light to collect data, which means that data can be collected during a 24 hour period as a frontal system develops.

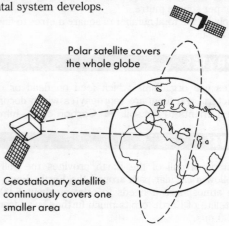

Fig S.5 Satellite types

SEASONS

The seasons are caused by the 23° tilt of the Earth's axis which means that different parts of the World get different amounts of sunlight. *Winter* occurs in the half of the Earth which is tilted away from the Sun; *summer* occurs in the half of the Earth which is tilted towards the Sun. When it is summer in the Northern hemisphere it is winter in the Southern hemisphere (Fig S.6).

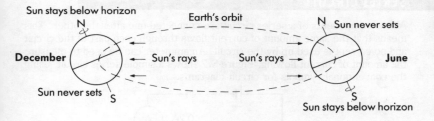

Fig S.6 How the seasons are caused

SECONDARY COLOURS

◀ Primary colours ▶

SEDIMENTARY ROCKS

Sedimentary rocks are formed when other rocks in the Earth's crust are broken down or eroded in one or more of the ways:

1 by the action of moving water;
2 by wind, ice and frost;
3 by changes of temperature;
4 by chemical action, such as acidic rainwater on limestone;
5 by the action of living organisms such as worms and plant roots.

The result of erosion is to break down the rocks into smaller particles and eventually form *soil*. Different soils are formed from different types of rocks so soils will differ in their drainage properties, texture, pH, and mineral composition. Rivers carry the fine particles such as gravels and sands, and deposit them on the sea bed, where they form successive layers over millions of years. As the layers become compressed, layers of sedimentary rocks or *strata* are formed, such as sandstone and limestone. You can sometimes see these strata at the coast where the layers may have been lifted up by movements of the Earth's crust.

SENSOR

A sensor is a receptor in a **feedback system** which detects any change from the normal. For example, in an oven the sensor is the **thermostat**. Other sensors are used to detect changes in levels of light intensity to switch on street lighting. You may have seen a 'smoke detector' which would detect high levels of smoke in a building.

◄ Control system ►

SERIES CIRCUIT

All the components of a series circuit are connected one after the other. This means that the same amount of current flows through each part of the circuit and no current is used up by the circuit. An ammeter can be used to measure the amount of current flowing. Figure S.7 shows a simple series circuit, using the conventional symbols for circuit diagrams.

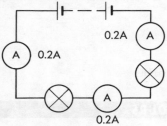

Fig S.7 Current in a series circuit

Note: as the charged particles go through each lamp they lose some of their energy, and so each lamp in the circuit becomes dimmer than the one before. The **voltage** measured across each lamp is equal to the total voltage measured across the battery.

SEXUAL REPRODUCTION

Sexual reproduction usually involves two parents, although there are many organisms which have both female and male sex organs. Such plants and animals are called *hermaphrodites* and can make both male and female gametes (sex cells). These organisms usually exchange **gametes** with others of the same species. Hermaphrodites include earthworms and buttercups. Sexual reproduction always involves the fusion of two gametes – one male sex cell (sperm) and one female sex cell (egg). The new cell produced, the **zygote**, divides many times to form an embryo and eventually grows to become the young organism. Two problems have to be solved by organisms reproducing in this way:

1 how to find a member of the opposite sex;
2 how to get the sperm and egg together.

Different ways of overcoming these problems have evolved. In humans, **fertilization** is *internal* so the egg and sperm fuse inside the female's body. In frogs, fertilization is *external* with the sperm being shed over the eggs which are laid in water. Usually offspring which are produced by internal fertilization are better protected as they develop either in an egg with a tough shell, or inside the mother as in humans.

The main advantage of sexual reproduction compared with asexual reproduction is that the offspring will vary from each other and from the parents – variety can mean the difference between success or extinction for the species in a constantly changing environment. ◄ Reproduction ►

SILVER

Silver is a metallic element, chemical symbol Ag. It is near the bottom of the **reactivity series**, so does not corrode easily and is used for electroplating other metals to protect them. It is quite rare, so is used as jewellery.

SKELETON

The function of your skeleton is to support your body and give muscles a firm attachment so that they can contract to move your bones (Fig S.8).

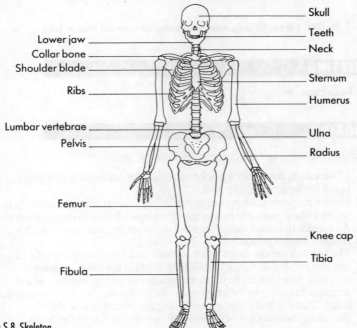

Fig S.8 Skeleton

SMELTING

Between your bones are *joints*; for example at your shoulder is a *ball and socket* joint which allows your arm to swivel around in any direction. At your elbow is a *hinge* joint which allows your lower arm to move backwards and forwards (Fig S.9).

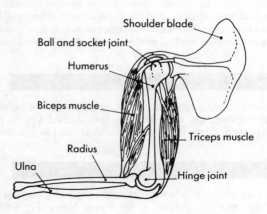

Fig S.9 The bones and muscles of the upper arm showing the elbow and shoulder joints

SMELTING

◄ Reduction ►

SMOKING

Cigarette smoking can affect the body in many ways:

1. Chemicals in the tobacco smoke can cause cancer in the lungs, and as a result the lungs are destroyed.
2. Carbon monoxide, a gas in cigarette smoke, mixes with haemoglobin in the **red blood cells** and makes the blood less efficient at carrying oxygen. As a result the blood vessels around the heart become weak and this may result in a heart attack.
3. The tiny hairs in the lungs which remove dust and mucus from the lungs become paralysed, so sticky *phlegm* collects in the lungs, causing infection. Smokers try to move the phlegm by heavy coughing which damages the lining of the lungs and reduces the number of air sacs in the lungs. There is less surface area for oxygen to diffuse into the blood stream so the smokers become out of breath and may suffer from bronchitis.

4 Pregnant women who smoke can give birth to babies which are undersized and sometimes born prematurely. ◀ Drugs – nicotine ▶

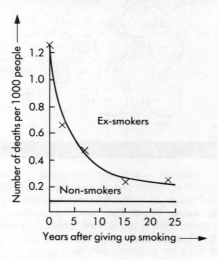

Fig S.10 Death rates and cigarette smoking

SODIUM CHLORIDE

Sodium chloride is *common salt*, the salt we use on our food. The chemical formula for sodium chloride is NaCl. Sodium chloride contains sodium ions (Na^+) and chloride ions (Cl^-) (Fig S.11a)). It can be found in sea water, and is often extracted from sea water by allowing the water to evaporate in large, shallow ponds leaving white crystalline salts (Fig S.11b)). The salts contain mostly sodium chloride but also other salts such as potassium chloride. In Britain, salt is found in large deposits in such places as Cheshire, where it is mined.

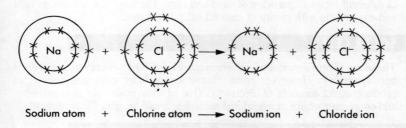

Sodium atom + Chlorine atom ⟶ Sodium ion + Chloride ion

Fig S.11 Sodium chloride   a) Ionic structure

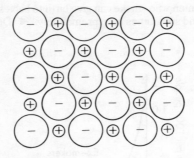

b) Crystalline structure

SOFT WATER

Tap water often contains dissolved salts, depending on what type of rock the rain water from which it comes has washed through. Different parts of the country will have different salts and different amounts of the salts will have dissolved in their water. **Water** that contains relatively large amounts of magnesium and calcium salts is called **hard water**. Soft water contains none, or very few, salts of magnesium or calcium (although it may contain other salts). Soft water will form a lather easily with soap.

SOIL

Soil is formed by the weathering of rocks which produce particles. These particles trap air and water and form a habitat for soil organisms such as earthworms, insects and bacteria.

Soils consist of mineral particles of varying size, water, air, mineral salts, humus, micro-organisms, insects and other soil animals.

The mineral particles determine the characteristics of the soil. For example, larger particles make up a *sandy* soil which is usually light, warm and easy to cultivate. Sandy soils can be easily **eroded** and mineral salts are quickly washed away (**leached**).

Clay soils are usually heavy, cold and difficult to dig. The small particles stick together and prevent drainage. Fertile, loamy soils are usually mixtures of different types of particles of sand and clay. The best soil, a loam, is light and easy to dig with plenty of mineral salts and water.

SOLAR CELLS

These are photoelectric devices which absorb the Sun's energy and convert it into electricity. However, many thousands of solar cells are needed to produce useful amounts of electricity. One of their main uses is in satellites and spacecraft where conventional batteries would be difficult to replace!

SOLAR ECLIPSE

◄ Eclipse ►

SOLAR ENERGY

Solar energy is the energy from the Sun. A common use of solar energy is to heat up water which is inside a solar panel. These panels are usually painted black so that they absorb as much heat as possible. The solar-heated water is then pumped to an ordinary hot water tank where the water can be heated electrically (Fig S.12). It is obviously much cheaper to heat water which has already been warmed up than to heat water from cold.

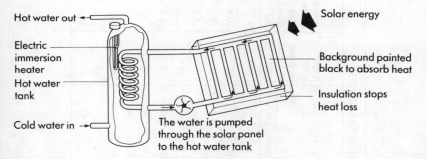

Fig S.12 Solar energy: a hot water system which uses solar energy

SOLAR SYSTEM

Our **solar system** consists of the Sun and nine known **planets** (of which Earth is one) which orbit around it. Figure S.13 shows the arrangement of the planets (this is not drawn to scale). Figure S.14 shows you some data about the nine planets in the Solar system.

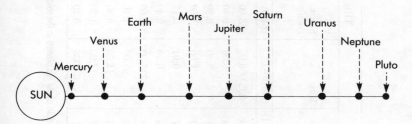

Fig S.13 Solar system: How the planets are arranged

In general there are two groups of planets. The planets nearer the Sun – Mercury, Venus, Earth and Mars, have small diameters and high density, while the planets further away – Jupiter, Saturn, Uranus, Neptune and Pluto, have large diameters but low density. Planets which are further away from the Sun take longer to orbit the Sun and generally have lower mean surface temperatures than those nearer the Sun.

SOLAR SYSTEM

THE MAIN MEMBERS OF THE SOLAR SYSTEM

Body	1 Diameter (Earth=1)	2 Mass (Earth=1)	3 Surface gravity (Earth=1)	4 Density, in Kg m^{-3}	5 Period of spin (days)	5 Period of spin (hours)	5 Period of spin (minutes)	6 Angle of tilt between axis and orbit	7 Average distance from Sun (Sun-Earth =1)	8 Period of orbit, in years	9 No. of moons (★ = plus rings)
Sun	109.00	333 000.00	28.00	1400	25	9					
Mercury	0.40	0.06	0.40	5400	58	16		97°	0.4	0.2	0
Venus	0.95	0.80	0.90	5200	244	7		90°	0.7	0.6	0
Earth	1.00	1.00	1.00	5500		23	56	267°	1.0	1.0	1
Moon	0.27	0.01	0.17	3300	27	7		113°	1.0	1.0	0
Mars	0.53	0.10	0.40	4000		24	37	91°	1.5	1.9	2
Jupiter	11.18	317.00	2.60	1300		9	50	114°	5.2	11.9	16★
Saturn	9.42	95.00	1.10	700		10	14	93°	9.5	29.5	15★
Uranus	3.84	14.50	0.90	1600		10	49	116°	19.2	84.0	5★
Neptune	3.93	17.20	1.20	2300		15	48	187°	30.1	164.8	2
Pluto	0.31	0.0025	0.20	400	6	9	17	118°	39.4	247.7	1

Fig S.14 Solar system: some data about the planets in the solar system

SOLENOID

◀ Electromagnetism ▶

SOLID

All matter can be classified as either solid, **liquid**, or **gas**. These are called the three *states of matter*. In a solid the particles are close together and are vibrating in 'fixed' positions. Solids keep their shape and resist being changed in shape. Some solids occur naturally with straight sides and sharp corners; these are called **crystals**.

SOLUBILITY

Substances will dissolve in **solvents** to varying degrees. The amount of the substance that will dissolve in any solvent is called its solubility.

The solubility of a **salt** often increases with temperature although there are a few exceptions. This is because for water to dissolve the salt it has to break down the *ionic lattice*. The higher the temperature, the greater the **kinetic energy** of the particles (water molecules and ions).

▶ SOLUBILITY OF SALTS

Cations

All salts containing sodium, potassium or ammonium ions are soluble.
Note: the ammonium ion (NH_4^+) is not a metal ion but may be regarded as such in salts.

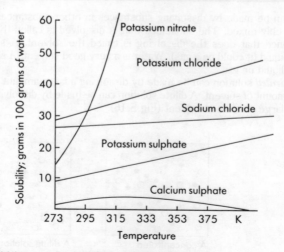

Fig S.15 Solubility curves; how solubility varies with temperature

Anions

- all nitrates (NO_3^-) are *soluble*;
- all chlorides (Cl^-) are *soluble*, except silver and lead;
- all sulphates (SO_4^{2-}) are *soluble*, except barium, calcium and lead;
- all carbonates (CO_3^{2-}) are *insoluble*, except sodium, potassium and ammonium.

If the solubility of a salt at different temperatures is plotted on a graph, *solubility curves* are obtained which show how solubility of a substance changes over a range of temperatures (Fig S.15).

SOLUTE

A solute is a substance that will dissolve in a **solvent** to produce a solution. Solutes can be solid, liquid or gas:

Solution	Solute	Solvent
lemonade	carbon dioxide	water
wine	alcohol	water
sea water	salts	water

Solid solutes can be divided into two groups:

- those which contain *ions* – these often dissolve in water.
- those which contain *molecules* – these are often insoluble in water but will dissolve in certain other solvents.

SOLUTION

Solutions can be made by dissolving substances in other substances so that they are evenly mixed. The substance that is dissolved is called the **solute**. The substance that does the dissolving is called the **solvent**. A solvent is normally a liquid (it could be a gas). Water is a very good solvent. A solute can be a solid, liquid or gas.

A *concentrated* solution can be made by dissolving a large amount of solute in a small amount of solvent. A dilute solution can be made by dissolving a little solute in a large amount of solvent (Fig S.16).

- Solute
- ○ Solvent

Fig S.16 Solutions

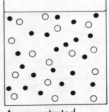

A concentrated solution

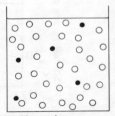

A dilute solution

The concentration of solutions can be measured as grammes of solute per cubic decimetre of solution (g/dm^3) or as moles of solute per cubic decimetre of solution (mol/dm^3). A *saturated* solution is one which contains the maximum amount of dissolved solute. ◄ Mole ►

SOLVENT

A solvent is a substance – usually a liquid (sometimes a gas) which can dissolve another substance to produce a solution.

Water is a good solvent since it can dissolve many substances e.g. salts, sugar, or gases such as oxygen and carbon dioxide. Some substances however are insoluble in water, but may dissolve in other liquids e.g. propanone will dissolve nail varnish, xylene will dissolve sulphur. Solvents such as these are non-aqueous solvents and are important for dissolving oils and grease. Perchloroethene is a solvent used in 'dry cleaning' – dry because water is not involved. A bottle of correction fluid contains a solvent 1,1,1-trichloroethane, and carries a hazard warning symbol (Fig S.17).

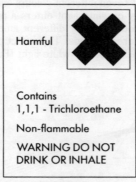

Fig S.17 Solvent: a hazard warning on a correction fluid bottle

SOUND WAVES

Sound waves are the only waves which are **longitudinal waves**. These waves are like the waves produced by a long spring as shown in Figure S.18. The energy is being transferred along the spring, but the particles are oscillating from left to right. ◄ Amplitude, frequency, speed, wavelength ►

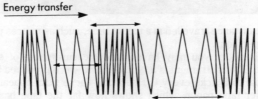

Fig S.18 Sound waves

SPECIES

This word describes organisms which are related closely enough to breed successfully. The offspring can reproduce themselves when they mature. It is possible for horses and donkeys to produce offspring called *mules*, but these are sterile and cannot breed. Of all the groups that we use to classify living things, the species (breeding group) is the smallest.

SPECIFIC LATENT HEAT

◀ Latent heat ▶

SPECTRUM

White light is made up of seven different colours each of which has a different **wavelength**. When a ray of white light enters a prism each of the different wavelengths are **refracted** (bent) by different amounts because they travel through the prism at different speeds. This effect produces a spectrum of all the different colours which make up white light (Fig S.19).

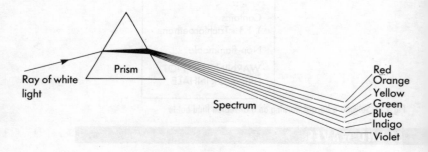

Fig S.19 Spectrum: White light is split into a spectrum of colours as it passes through the prism

SPEED

Speed is the distance travelled in a unit of time, such as metres per second, or kilometres per hour.

In the laboratory you may have made measurements of speed using a ticker-timer. This instrument is a type of clock which produces 50 ticks every second, equal to 5 ticks every 0.1 second. These ticks appear as dots on a strip of ticker tape paper, which shows how far the tape has been moved between each dot. The time interval between each dot is 0.02 seconds. When the dots are close together the tape has been moved slowly. When the dots are far apart the tape has been moved quickly (Fig S.20).

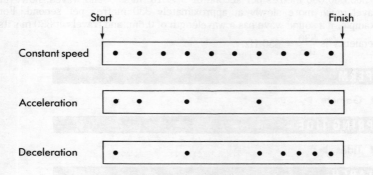

Fig S.20 Speed; ticker tapes

The ticker tape is usually attached to a moving trolley to study how the trolley moved. The tape showing in Figure S.21 was produced by the trolley moving down the runway. The trolley started slowly and then accelerated.

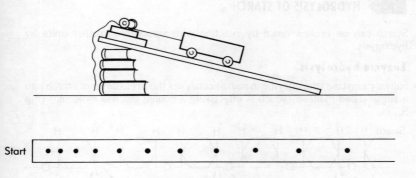

Fig S.21 Speed; the ticker tape shows how the trolley accelerated

SPEED, FREQUENCY AND WAVELENGTH

Speed (metres per second) = frequency (hertz) × wavelength (metres)

$$v = F \times \lambda$$

For example, if a wave is travelling with a frequency of 30 Hz and has a wavelength of 3 m, its velocity is 90 m/s.

The waves in the electromagnetic spectrum all travel at the same velocity of 300,000,000 metres per second or 3×10^8 m/s. Sound waves, however travel much more slowly at approximately 330 metres per second. For example, if a sound wave has a wavelength of 0.6m, and travels at 330 m/s its frequency is $\frac{330}{0.6} = 550$ Hz.

SPERM

◀ Gamete ▶

SPRING TIDE

◀ Tides ▶

STARCH

Starch is a natural **polymer** which is made by green plants. It consists of long chains of glucose molecules (the monomer) joined together. Starch is a **carbohydrate** and as a food is a useful source of energy. The starch molecule is too large to pass through the gut wall into the blood system, so is broken down into smaller units in the digestive system.

▶ HYDROLYSIS OF STARCH

Starch can be broken down by reaction with water into smaller units by hydrolysis.

Enzyme hydrolysis

Salivary amylase (the enzyme in saliva) catalyses the breakdown of starch into a sugar called maltose (which is effectively a double glucose molecule) (Fig S.22).

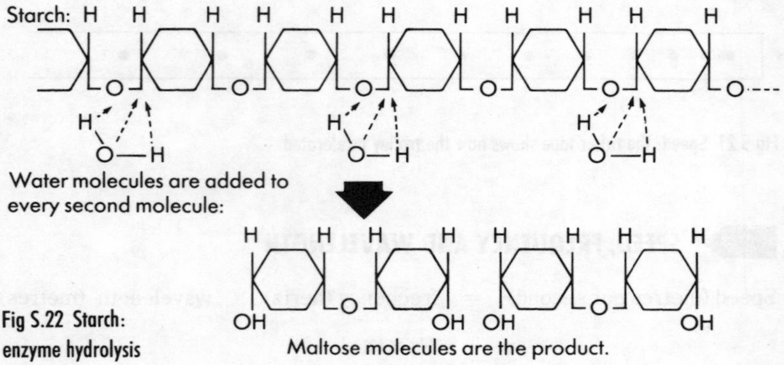

Fig S.22 Starch: enzyme hydrolysis

Water molecules are added to every second molecule:

Maltose molecules are the product.

Acid hydrolysis

Acids can also catalyse the breakdown of starch but the product is glucose. The stomach contains hydrochloric acid which can help break down any unconverted starch or maltose molecules into glucose. Glucose molecules are small enough to pass through the gut wall.

STATE SYMBOLS

Chemical equations tell us which substances react and what products are formed. They can also tell us the *state* of the reactants or products, if for example they are in a gaseous form or if they are dissolved in water. This is important since some reactants will only react if they are in a particular state. We can show the state of reactants and products by adding symbols of state:

- (g) means gas
- (l) means liquid
- (s) means solid
- (aq) means dissolved in water

Examples of using these symbols are as follows:

$$2Na(s) + Cl_2(g) \rightarrow 2NaCl(s)$$

$$Mg(s) + 2HCl(aq) \rightarrow MgCl_2(aq) + H_2(g)$$

HCl (aq) indicates that *dilute* hydrochloric acid is being used.

STATES OF MATTER

◀ Solid, liquid, gas ▶

STEEL

Steel is an **alloy** of the metal iron mixed with a small quantity of carbon (a non-metal). The introduction of a small quantity of carbon increases the strength of the iron enormously because the carbon atoms fit into the giant structure of iron atoms, preventing them moving so freely when the material is hammered, twisted or stretched (Fig S.23).

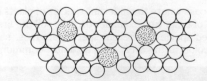

Fig S.23 Steel structure: atoms of carbon are mixed with atoms of iron

Different steels can be produced with different properties, depending on the amount of carbon present. Other metals can also be added, such as chromium or tungsten to give steel even more properties. These different types of steel are used for various purposes.

Steel	Use	Constituents	Important properties
spring steel	suspension springs	iron 0.3%–0.6% carbon	contains sufficient carbon that will produce a springy metal
stainless steel	surgical instruments; cutlery	iron <1% carbon 18% chromium	resistant to corrosion
chromium-vanadium steel	axles and wrenches	iron chromium vanadium and carbon	very strong, great resistance to strain
high-speed tungsten steels	cutting metals, drills etc.	iron with up to 20% tungsten	maintains sharp edge at high temperatures

STEP-DOWN TRANSFORMER

◄ Transformer ►

STEP-UP TRANSFORMER

◄ Transformer ►

STRATOSPHERE

◄ Atmosphere ►

SUBLIMATION

◄ Iodine ►

SUCCESSION

Succession is the process where the different species of plants in a **community** change over a period of time. For example, on bare soil the first plant species to arrive may be weed species. Different species of grass, shrubs and trees will then eventually colonise the area, until a **climax community** is reached.

SUGARS

Sugars are **carbohydrates**: sugar molecules consist of carbon, hydrogen and oxygen atoms. Simple sugars are manufactured by plants during the process of **photosynthesis**. There are different types of sugars; the names of sugars all end in -ose.

SULPHURIC ACID

- glucose: chemical formula $C_6H_{12}O_6$ – the simple sugar produced by plants during photosynthesis.
- sucrose: chemical formula $C_{12}H_{22}O_{11}$ – the sugar we use to sweeten tea, found in large amounts in plants such as sugar cane; sucrose is like a double glucose molecule.
- lactose: the sugar found in milk.
- fructose: the sugar found in fruit.
- maltose: found in malted barley.

SULPHATES

Sulphates are **salts** which contain the sulphate ion (SO_4^{2-}). Examples are copper sulphate, magnesium sulphate, calcium sulphate. Most sulphates are soluble in water, except those of calcium, barium and lead. Calcium sulphate is found as the rock *gypsum* and is used for making plaster.

SULPHUR DIOXIDE

Sulphur dioxide (SO_2) is a poisonous, choking gas. Sulphur dioxide is formed when sulphur is burned in air:

sulphur + oxygen → sulphur dioxide
$S(s)$ + $O_2(g)$ → $SO_2(g)$

Since sulphur is present as an impurity in fuels such as coal and oil that are burned to generate electricity, sulphur dioxide can be released into the atmosphere as a *pollutant*.

When metals are extracted from metal sulphide ores, the ore is first roasted to produce a metal oxide, and sulphur dioxide is a waste product:

lead sulphide + oxygen → lead oxide + sulphur dioxide

Sulphur dioxide is a major contributor to **acid rain** since it reacts with oxygen and ozone in the atmosphere to produce **sulphuric acid**.

SULPHURIC ACID

Sulphuric acid (chemical formula H_2SO_4) is a strong **acid** which is very important to the chemical industry since it is involved with the manufacture of many other chemicals.

Manufacture of sulphuric acid

The raw materials for the manufacture of sulphuric acid are sulphur dioxide gas and oxygen from the air. The sulphur dioxide gas can come from burning sulphur in air or the roasting of metal sulphide ores.

Sulphur dioxide is converted into sulphur trioxide by mixing it with oxygen (air) and passing the mixture through a catalyst – vanadium (V) oxide:

$$2SO_2 + O_2 \underset{\text{oxide catalyst}}{\overset{\text{Vanadium}}{\rightleftharpoons}} 2SO_3$$

The sulphur trioxide reacts with water to produce sulphuric acid. Because this can produce dangerous 'clouds' of acid mist, the SO_3 is dissolved in concentrated sulphuric acid, then water is added.

Sulphuric acid as an acid

Dilute sulphuric acid will change the colour of **indicators**, will react with **metals** to form salts releasing hydrogen, and will react with **bases** to form salts and water.

Concentrated sulphuric acid is also a good *dehydrating agent* (removes water molecules from some compounds).

Uses of sulphuric acid

The manufacture of fertilizers, paints, detergents, plastics, fibres, metals and alloys, battery acid (Fig S.24).

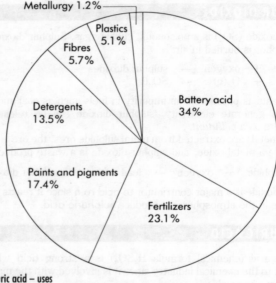

Fig S.24 Sulphuric acid – uses

SUN

The Sun is made of hydrogen and helium gas; the temperature is about 6000°C at its surface. At the centre of the Sun the temperature is much higher and this is where **nuclear fusion** is taking place as the hydrogen is being converted to helium. The Sun is about half way through its life cycle of 9,600 million years and is about a million times larger than **Earth**.

The Sun is one of billions of stars which make up a galaxy known as the Milky Way and there are billions of galaxies like the Milky Way in the universe. Figure S.25 shows the life cycle of a *star* such as the Sun.

SYMBOLS

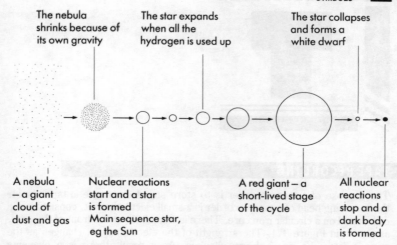

Fig S.25 The Sun: the life cycle of a typical star

SURFACE COATING

◀ Corrosion ▶

SYMBIOSIS

This occurs when two organisms live together and are useful to each other. For example, nitrogen-fixing bacteria which live in the root nodules of beans and peas. The bacteria gain carbohydrates from the plant, and the plant gains by using the nitrates which the bacteria have made by fixing atmospheric nitrogen. ◀ Nitrogen cycle ▶

SYMBOLS – ELECTRICAL

◀ Circuit symbols ▶

SYMBOLS – STATE

◀ State symbols ▶

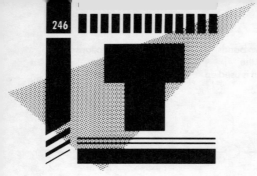

TAPE RECORDING

The purpose of a tape recorder is to store sound on magnetic tape. Inside the recording head of a tape recorder is a small electromagnet, consisting of a coil wound on a circular iron core. There is a small slit at the front of the core, as shown in Figure T.1. The strength of the electromagnet changes as the current from the **microphone** changes. As a result there is a changing magnetic field across the slit, which magnetises the particles on the tape coated with iron oxide or chromium oxide. The pattern of particles on the tape therefore reflects the pattern of the changing strength and frequency of the sound waves. When the magnetised tape passes the *playback* head then the magnetic patterns are converted back into sound.

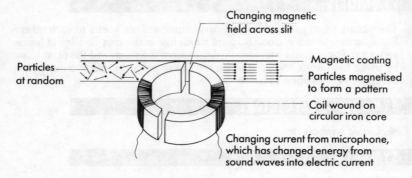

Fig T.1 Tape recording: the recording head is an electromagnet

TELEVISION PICTURES

Colour pictures on a television screen are produced by the following process:

1. When a TV scene is filmed, light enters the camera, and is split into red, blue and green light;
2. these different signals then go to three different 'tubes' which then send out the corresponding signal on ultra-high frequency waves, to the three *electron guns* at the back of the TV screen;

3 the electrons hit millions of tiny red, green and blue dots covering the TV screen which glow when the electrons hit them;
4 depending on which coloured dots glow, the different colours are produced on the screen.

TESTIS

The male sex organ which produces sperm, or male sex cells (**gametes**). The testes are usually held in the *scrotum*, a bag of skin which hangs outside the body where the temperature is slightly lower than that inside the body. The sperm develop better at a lower temperature. Inside the testis is a 50 cm long, narrow tube where the sperm are produced by **meiosis**.
◄ Reproduction ►

THERMOSETTING PLASTICS

These are **plastics** which can be heated and moulded only once; when they cool and harden they cannot be remoulded by heating. They are manufactured in two stages. First, a resin is produced of long chains. Second, this resin is heated and moulded, so that cross-links form between the molecules. The cross-links are strong chemical bonds which hold the shape rigid, and are not broken down by reheating (Fig T.2).

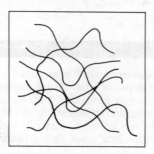

Fig T.2 Thermosetting plastics; molecular structure

Examples of thermosetting plastics are bakelite, epoxy resin and melamine-formaldehyde. ◄ Plastics ►

THERMOSOFTENING PLASTICS

These are **plastics** which can be softened when heated and harden again when they cool down. This can be repeated many times. *Nylon*, for example, can be melted and extruded through tiny holes to produce a fibre, or moulded to produce an object of any desired shape.

The plastic molecules are held together by weak inter-molecular forces. Heating provides the molecules with more energy so they slip past each other more easily. On cooling, they lose this extra energy and the weak forces of attraction take over again, so the plastic becomes rigid (Fig T.3).

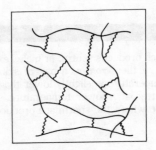

Fig T.3 Thermosoftening plastics; molecular structure

Thermosoftening plastics can be made of different densities, e.g. *high density* or *low density* polythene. As the name suggests, high density polythene has its molecules much closer together, lying almost parallel to each other. This type of polythene will stand up to more wear and tear and softens at a higher temperature.

Examples of thermosoftening plastics are nylon, polythene, polypropylene, PVC and polystyrene. ◀ Plastics ▶

THERMOSTATS

Thermostats are control devices which keep a fluid or an object at a constant temperature. **Bimetallic strips** are often used in thermostats. These are made up from two dissimilar metals joined together, for example brass and iron. When heated, the metals expand at different rates and cause the strip to bend. This bending movement can be used to switch various devices on or off. Thermostats are used in domestic irons, central heating systems, fire alarms, ovens etc. (Fig T.4).

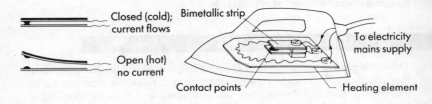

Fig T.4 Thermostat: bimetallic strip thermostat in an iron

THREE-PIN PLUG

Figure T.5 shows the correct connections for a three-pin plug:

1. The *live* wire is coloured brown, and is connected to the live pin.
2. The *neutral* wire is coloured blue, and is connected to the neutral pin.
3. The *earth* wire is coloured green and yellow, and is connected to the earth pin at the top of the plug.

◀ Fuse, earth wire ▶

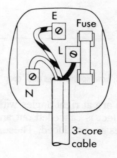

Fig T.5 Three-pin plug; how to connect a three core cable

TIDAL POWER

Tidal power involves using the massive changes in the level of the sea which occur twice a day as a result of the gravitational effect of the Sun and Moon on the Earth. This causes regular tidal movements of the oceans which are used to push water into reservoirs, from where the water is used to drive turbines and produce electricity. There is a very large power station on the coast of Brittany in France which uses tidal power.

TIDES

The **Moon** exerts a gravitational pull on the water which is on the surface of the **Earth**. The effect of this, on the side *nearest* the Moon, is to pull the water towards the Moon and produce a high tide. A high tide also happens on the side of the Earth *furthest away* from the Moon. Due to the rotation of the Earth these high tides happen every 12 hours.

Although the **Sun** is much further away from the Earth, it too has a gravitational pull on the Earth (Fig T.6). When the Sun and Moon are in line, about twice a month, their combined gravitational pull causes a very high tide or 'spring' tide. These tides have very large tidal ranges, this means they have 'very high' high tides and 'very low' low tides.

TITRATION

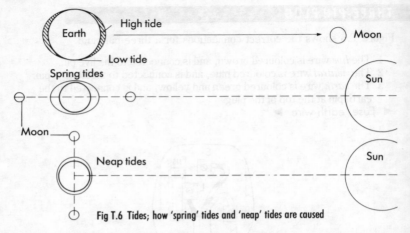

Fig T.6 Tides; how 'spring' tides and 'neap' tides are caused

When the Sun and Moon are at right angles with each other, about twice a month, the gravitational effect is cancelled out and weak tides, called 'neap' tides, with small tidal ranges are produced. These tides have 'low' high tides.

TITRATION

◄ Indicators ►

TRANSFORMER

There are three important facts to remember about transformers:

1. transformers change voltage;
2. transformers only work on alternating current;
3. transformers contain an iron core and two coils of wire, a primary coil and a secondary coil.

A *step-up transformer* gives out a higher voltage than the input voltage and has more turns on the secondary coil than the primary coil (Fig T.7a)).

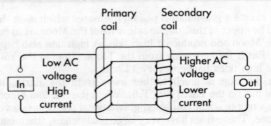

Fig T.7 Transformers a) Step-up transformer

A *step-down transformer* gives out a lower voltage than the input voltage, so there are more turns on the primary coil than the secondary coil (Fig T.7 b)).

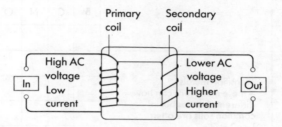

b) Step-down transformer

When the primary coil is connected to an alternating current, it acts like an electromagnet which is switched on and off very quickly, causing a current to flow backwards and forwards. This sets up a changing magnetic field in the iron core which induces an alternating current in the secondary coil. You can calculate the voltage induced in the secondary coil using the following formula:

$$\frac{\text{voltage across secondary coil}}{\text{voltage across primary coil}} = \frac{\text{number of turns in secondary coil}}{\text{number of turns in primary coil}}$$

In symbols, $\frac{V_2}{V_1} = \frac{N_2}{N_1}$

For example, if a step down transformer has 100 turns on the primary coil and 10 turns on the secondary coil, you can calculate the output voltage given that the input voltage is 240 V:

$$V_2 = \frac{10}{100} \times 240 = 24V$$

The output voltage is 24 volts.

One of the main uses of transformers is in the National Grid system where step-up transformers increase the voltage and lower the current so that less electricity is wasted as heat. Step-down transformers are used in many household appliances such as televisions, computers, radios and washing machines in order to reduce the mains voltage to a lower voltage.

TRANSITION METALS

The transition metals are found in a block in the **periodic table** (Fig T.8). They include those we use and see every day such as copper, zinc, iron, silver etc. When these metals react they can form ions with different charges (they have variable valency); the charge on their ions is usually 1+, 2+, or 3+. For example, copper can form ions of Cu^+ or Cu^{2+}; iron can form ions of Fe^{2+} or Fe^{3+}.

Transition metals in the periodic table

Periods			Groups					O
I	II		III	IV	V	VI	VII	He
Li	Be		B	C	N	O	F	
		Transition metals — these metal ions can have different charges but in common salts are often 1+, 2+, 3+						

Fig T.8 Transition metals in the periodic table

The *size* of the charge on the ion in a compound is shown by roman numerals. For example, in copper (II) oxide, the ion present is Cu^{2+}; in copper (I) oxide, the ion present is Cu^{+}.

The compounds of transition metals are also often coloured (Fig T.9).

Compound	Colour	Formula	Metal ion
Copper (II) sulphate	blue	$CuSO_4$	Cu^{2+}
Iron (II) sulphate	green	$FeSO_4$	Fe^{2+}
Iron (III) oxide	red	Fe_2O_3	Fe^{3+}
Copper (I) oxide	red	Cu_2O	Cu^{+}
Copper (II) oxide	black	CuO	Cu^{2+}

Fig T.9 Transition metals: colours of the compounds

TRANSMISSION OF ELECTRICITY

Electricity from a power station is transmitted across the country by the National Grid system. The commonest method is by overhead power cables, carried on pylons. Sometimes, underground transmission lines are used. Each method has its advantages and disadvantages:

- Overhead cables: these are cheaper to install, and easier to repair, but unsightly, dangerous to people (especially those using kites), and to people moving boats with high masts or when carrying fishing rods.
- Underground cables: these are more expensive to install, and more difficult to repair, but can be hidden underground and present no danger to people.

Electricity is transmitted from power stations at voltages of 400,000 V. The reason for using such high voltages is that there is a very low current and energy loss is very small. If electricity was transmitted at a lower voltage there would be a greater current and more energy would be lost as heat (Fig T.10).

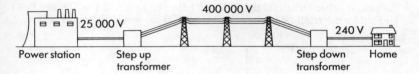

Fig T.10 Transmission of electricity: how electricity gets from the power station to your home

TRANSPIRATION

This is the loss of **water vapour** through the *stomata* of plant leaves. The rate of transpiration can be increased by atmospheric conditions such as wind, high temperatures, low humidity. Leaves with a greater surface area or more stomata transpire more quickly.

TRANSVERSE WAVE

Electromagnetic waves are described as transverse waves because the particles move up and down at right angles to the direction of the wave (Fig T.11).

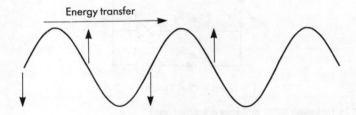

Fig T.11 Transverse wave: energy is being transferred along this transverse wave, but the particles only move up and down

TROPHIC LEVEL

◀ Producers, herbivores, carnivores ▶

TROPOSPHERE

◀ Atmosphere ▶

TRUTH TABLES

The common way of showing how circuits behave is to produce a truth table, which is a summary of what a circuit can do. The circuit in Figure T.12a) has the switch either on or off, and the lamp either on or off. If 1 = ON and 0 = OFF, then the truth table shown in Figure T.12b) would be produced.

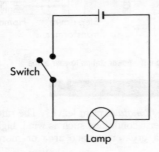

Switch	Lamp
0	0
1	1

a) On/off logic circuit

b) Truth table for on/off logic circuit

Fig T.12 Truth tables

The working of a more complicated circuit, shown in Figure T.13a) can also be summarised in a truth table (Fig T.13b)).

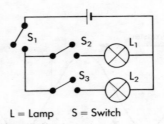

L = Lamp   S = Switch

Fig T.13 Truth tables   a) A more complicated circuit

b) Truth table for a more complicated circuit

Switch			Lamp	
S_1	S_2	S_3	L_1	L_2
0	0	0	0	0
1	0	0	0	0
1	1	0	1	0
1	0	1	0	1
1	1	1	1	1

TRUTH TABLES

As systems become more complicated they can be simplified by removal of the supply, with just two lines drawn to represent the positive [+] and negative [−] sides of the circuit (Fig T.14a). In these logic circuits the switch can be left on so that all possible connections can be made. The truth table (Fig T.14b) shows a summary of the circuit. ◄ Logic gates ►

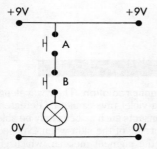

Switch		Lamp
A	B	
0	0	0
1	0	0
0	1	0
1	1	1

a) Series logic with two switches
b) Truth table for series logic circuit

Fig T.14 Truth tables

ULTRA-VIOLET RAYS

Ultra-violet rays are a form of **electromagnetic radiation**. Their wavelengths are shorter than those of visible light. Ultra-violet rays cannot be detected by the human **eye** but it is thought that some insects such as bees may be able to detect the rays. Ultra-violet rays cause burning of the skin when exposed to sunshine and causes the skin to produce the pigment melanin, which gives light-skinned people a suntan. Prolonged exposure to ultra-violet rays may cause harmful skin cancers, especially in fair-skinned people.

A use of ultra-violet rays is in the security marking of electrical equipment and other valuable objects. The marking only shows up under ultra-violet light so that stolen goods can be identified and returned to their owners.

UNDERGROUND CABLES

◀ Transmission of electricity ▶

UNIVERSAL INDICATOR

Indicators can be used to detect the difference between acids and alkalis. The most commonly used indicator is *universal indicator* since it not only tells us if something is **acid** or **alkaline** but also if the acid (or alkali) is strong or weak. It can be used as a liquid (usually green) or soaked onto a type of blotting paper and used as a paper. The paper has to be wet in order to work since acids only behave as acids in solution. The colour of the indicator matches a number which indicates the acidity of the solution.

The *strength* of an acid or alkali is measured on the **pH scale**. This has a range of 1 to 14 and is a measure of the hydrogen ion concentration (acidity). Notice that *low* numbers indicate *high* acidity (and high H^+ ion concentration) (Fig U.1). If a solution turns the indicator light green then it is *neutral* and has a pH of 7. If a solution turns the indicator yellow then it is a *weak acid* and has a pH of 6.

Following an acid/alkali reaction

Indicators can also be used to follow the course of a **neutralization** reaction between an acid and an alkali. If universal indicator is added to a strong alkaline solution the indicator will turn violet. If a solution of acid is added a small amount at a time then the indicator will change colour through blue to green at which point the solution is neutral (the acid has reacted with all the alkali present). If acid is continued to be added the indicator will eventually turn red indicating the solution is now strongly acidic.

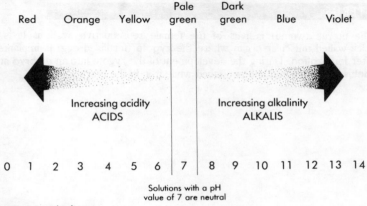

Fig U.1 Universal indicator

URANIUM

Uranium is a metallic element – chemical symbol U. The main source of uranium is pitchblende, in which it is found as an oxide. Uranium has three **isotopes** – ^{238}U, ^{235}U and ^{234}U.

^{235}U is of great importance since it is used as a nuclear fuel. When bombarded with neutrons it breaks into smaller pieces, releasing vast amounts of energy:

$$^{235}_{92}\text{U} + n \rightarrow {}^{144}_{56}\text{Ba} + {}^{90}_{36}\text{Kr} + 2n + \text{Energy}$$

All the isotopes of uranium are radioactive, ^{238}U decays to produce ^{234}U, but is not used as a nuclear fuel.

UREA

Urea is a waste product produced by the breakdown of excess **amino acids** in the liver, and removed from the body by the **kidneys**. It passes out of the body in **urine**. Adults excrete about 30g a day. Urea is a white, crystalline solid which is manufactured industrially for preparing urea-formaldehyde plastics, barbiturates (drugs) and as a fertilizer.

URINE

Urine consists mainly of water, with some dissolved salts and **urea**. It is excreted from the **kidneys** and stored in the bladder until being passed out of the body via the ureter. A doctor will sometimes test your urine to see if there are any substances present which may indicate that there is something wrong with you. For example, if there is sugar present in the urine it may indicate that a person is diabetic and in need of medical treatment.

UTERUS

The uterus (womb) is part of the female reproductive system. It is a thick-walled muscular organ where the **zygote** (fertilized egg) is implanted after **fertilization**. During the development of the zygote into an **embryo** and **foetus** it is protected and nourished while it is in the uterus.

VACCINE

A vaccine usually contains weakened or dead micro-organisms which are injected into the body to help overcome an infectious disease.

VALENCY

A formula for a compound shows the ratio of atoms present in that compound and whether they are joined by ionic or covalent bonds. Each atom has a 'combining power' which is called the *valency*. The valency of an atom depends on the number of electrons in its outer shell and hence its position in the periodic table. For example, atoms in Group 1 have a valency of 1; atoms in Group 2 have a valency of 2.

In general as one moves across the **periodic table** the valency gradually increases to a maximum of 4, then gradually decreases to 0, although there are some important exceptions.

	Na	Mg	Al	Si	P	S	Cl	Ar
Outer shell electrons	1	2	3	4	5	6	7	8
Group no	1	2	3	4	5	6	7	0
Valency	1	2	3	4	3	2	1	0

When atoms react to form ions, one atom has to *lose* electrons whereas the other atom has to *gain* electrons. Non-metal atoms like sulphur, which have 6 electrons in their outer shell, can be considered to have 2 spaces (to complete the full set of 8). It is easier to fill two spaces than it is to remove 6 electrons.

When atoms join to form molecules the number of electrons leaving one atom must equal that being gained by the other. It may help to imagine the atoms with hooks representing their valencies:

Example 1

Sodium will react with chlorine to form a compound, sodium chloride

Na has a valency of 1 (Na)⌒

Cl has a valency of 1: we can represent it as (Cl)⌒

When these atoms combine *all* hooks must be attached:

So the formula is NaCl.

Example 2

Magnesium will react with chlorine to form magnesium chloride

Magnesium: valency 2: (Mg)
Chlorine: valency 1: (Cl)

When they join, *all* hooks must be attached, so we need an extra Cl to take care of the otherwise spare hook:

The formula is therefore $MgCl_2$. The 2 as subscript refers to 2 atoms of what is immediately in front, i.e. Cl atoms.

Example 3

The formula of aluminium oxide

Aluminium: valency 3 (Al)
Oxygen: valency 2 (O)

When they join, *all* hooks must be attached:

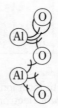

So the formula is Al_2O_3.

Sometimes we can regard a collection of atoms, referred to as *radicals*, as having a valency. For example:

sulphate	SO_4^{2-}	:	valency 2
nitrate	NO_3^{-}	:	valency 1
carbonate	CO_3^{2-}	:	valency 2
hydroxide	OH^{-}	:	valency 1

Example 4

The formula of copper nitrate.

Copper: valency 2 (Cu)

Nitrate: valency 1 (NO$_3^-$)

Formula Cu(NO$_3$)$_2$.

Notice the use of brackets with the 2 as subscript outside. This means 2 of whatever is inside the brackets.

You will not be expected to remember all the valencies for these atoms or radicals, but it is worth remembering how they are related to the *position* in the periodic table.

VARIATION

Organisms differ in their appearance even within the same family. These differences (variations) are the result of the new genes (called **mutations**) and new mixtures of **genes** (produced during sexual reproduction).

There are two types of variation between individuals of the same **species**:

1 *Discontinuous variation*. This is a marked change from one characteristic to another and enables us to separate individuals into distinct groups. One of the most used examples is blood grouping (Fig V.1). There are four main groups: A, B, AB and O. The information in the genes accounts for most of this form of variation and the environment affects it very little.

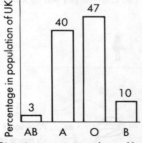

Fig V.1 Discontinuous variation in human blood groups

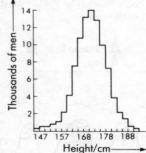

Fig V.2 Continuous variation in humans: height

2 *Continuous variation*. This refers to characteristics which gradually change within a population. We cannot separate individuals into distinct groups. Your height and weight are good examples of this sort of characteristic (Fig V.2). Many genes may influence height and weight, but the environment can be important also. In any large population you would get a whole range of heights and weights.

VASOCONSTRICTION

This is the narrowing or constriction of blood vessels, especially those just below the surface of the skin (Fig V.3). This results in heat being retained in the body, and is used in body temperature regulation.

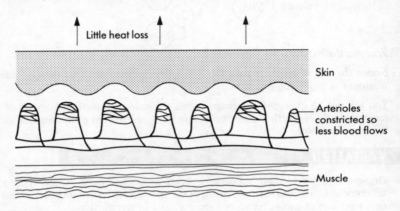

Fig V.3 Vaso-constriction

VASODILATION

This is the dilation or widening of blood vessels especially those just below the surface of the skin with the result that heat is lost from the blood and the person cools down (Fig V.4). Vasodilation can cause a person to look flushed and red in the face.

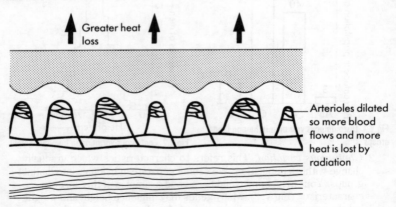

Fig V.4 Vasodilation

VEIN

Veins are large thin-walled blood vessels which carry blood towards the **heart** (Fig V.5). With the exception of the **pulmonary vein** the veins carry deoxygenated blood from the rest of the body. The veins have valves which allow the blood to flow in one direction only.

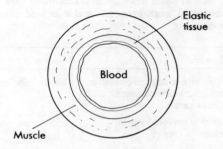

Fig V.5 Vein; veins have much thinner walls than arteries

VELOCITY

Velocity is the distance travelled in a certain direction in unit time. Velocity is a vector quantity because it has direction, whereas **speed** is a scalar quantity.

$$\text{velocity} = \frac{\text{distance moved in a stated direction}}{\text{time taken}}$$

The units of velocity are m/s (metres per second) or km/h (kilometres per hour) (Fig V.6).

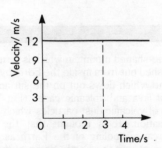

Fig V.6 Velocity: a graph showing constant velocity

VIRUS

◀ Micro-organism ▶

VISKING TUBING

◀ Osmosis ▶

VITAMINS

Vitamins are required in small amounts in a **balanced diet**. They are often necessary for *enzymes* to work properly, and a lack of vitamins in the diet causes deficiency diseases.

Vitamin	Benefits	Source
A	resistance to infection; vision in dim light; health of mucus membranes	butter, milk, carrots, liver
B	health of nervous system; release of energy	liver, yeast, wheat
C	health of blood vessels and skin	citrus fruit, potatoes, green vegetables
D	good bone development	fish liver oil, butter, milk action of sunlight in skin

A lack of vitamins often causes health problems:

- A lack of vitamin A causes poor 'night' vision, and can result in blindness in many less developed countries.
- A lack of vitamin B causes *beri-beri* (a type of paralysis).
- A lack of vitamin C causes *scurvy*.
- A lack of vitamin D causes *rickets* (poor bone formation resulting in 'bandy' legs).

VOLCANO

A volcano is a large cone-shaped mountain which is formed when steam, lava, rocks and gases are pushed out from inside the **Earth** by the pressure of gases and steam. The **magma** which flows out on the surface at a temperature of 1000°C is a mixture of lava and volcanic gases (Fig V.7). Some eruptions produce large amounts of volcanic dust particles which enter the atmosphere and cause cloud formation. It is thought that the volcanic ash in the **atmosphere** may also cause cooling of the Earth as the particles prevent radiation from the **Sun** reaching the Earth. Some of the effects of volcanoes are the destruction of towns and villages, and the removal of agricultural land and forests. A major volcanic eruption took place in 1976 in China where over 1 million people were killed. On the beneficial side: volcanic ash forms a very fertile soil; and molten rock underground heats underground water forming steam which can be used to generate **electricity**.

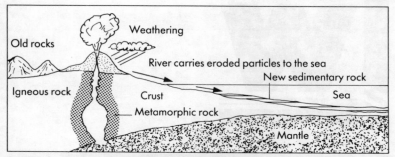

Fig V.7 Volcanoes: how new rocks are formed from volcanic eruptions

VOLT

A volt, symbol V, is the SI unit of electric potential, **potential difference** or **electromotive force**. One volt is defined as the potential difference between two points in an electric circuit if one **joule** of work is done, transferring one **coulomb** of charge between the points.

VOLTAGE

The voltage is the **potential difference** between two points of a circuit measured in **volts**. ◀ **Electromotive force, cathode ray oscilloscope** ▶

VOLTMETER

A voltmeter is placed in a circuit in **parallel** with the place where energy is being converted. For example in a lamp, energy is being converted into heat and light. When the voltmeter measures the change of electrical energy into another form of energy the voltage is described as the 'potential difference' or PD between the two points of a circuit. ◀ **Cathode ray oscilloscope** ▶

WATER

Water is a liquid – chemical formula H_2O. It freezes at 0°C and boils at 100°C. Water has its maximum density at 4°C while still liquid, rather than in its solid form as ice. This is why ice floats on water. Water can dissolve many different substances; it has been nicknamed 'the universal solvent'.

WATER AS A SOLVENT

In the past it was considered that 'like dissolves like'. In other words, liquids containing **ions** will often dissolve other ionic substances, while liquids that are molecular (contain **covalent bonds**) will often dissolve other covalent substances.

Water is an oddity, since it consists of water molecules in which the atoms are *covalently* bonded, yet it dissolves *ionic* substances. This is because the water molecule is *polar* (it has a slight negative charge ($\vartheta-$) at one end and a slight positive charge ($\vartheta+$) at the other (Fig W.1). This is because the oxygen atom attracts the electron pairs of the bonds formed with the hydrogen atoms more strongly. Water can therefore dissolve ionic substances. Water will also dissolve molecular compounds which are polar, e.g. sugar and ethanol.

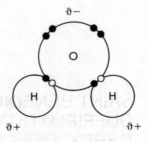

Fig W.1 Water: the polar water molecule

WATER CYCLE

Water in the ocean is continuously being evaporated by the heat of the Sun, and this vapour condenses to form **clouds**. When the clouds are blown over hills and mountains, they release the condensation as *rain*. Some of the water drains through the ground and back to the sea by rivers, and some of it is absorbed through the roots of plants, and is evaporated from the leaves in the process of transpiration (Fig W.2). Rain also dissolves some of the poisonous gases in the air, such as sulphur dioxide, and forms dilute sulphuric acid, which falls as **acid rain**.

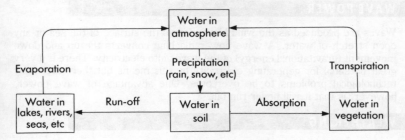

Fig W.2 Water cycle

WATER VAPOUR

Water vapour is present in the **air** to varying degrees, depending on the temperature of the air, land temperature, amount of wind etc. Water vapour is not visible, however clouds are caused by water vapour condensing into small droplets of water in the colder air above the Earth. Water vapour gets into the air as a result of the Sun evaporating water from the seas, rivers and lakes.

WATT

The watt is a unit of power. A power of one watt is produced when 1 joule of work is done in 1 second. The symbol for the watt is W. A useful formula to remember is watts = volts × amps.

WAVELENGTH

The wavelength is the distance between one point of a wave and the next point at the same place, as shown in Figure W.3. For example, the distance between the crest of one wave and the next. The symbol for wavelength is λ. Wavelength is equal to the speed of the wave, V, divided by its frequency, f.

$$\lambda = \frac{V}{f}$$

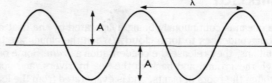

Fig W.3 Wavelength

◀ Amplitude, speed, frequency, wavelength ▶

WAVE POWER

Waves are produced as the wind blows across the surface of the sea or any open stretch of water. A wave power machine converts the up and down movement (gravitational energy) of the waves into electricity. There is a very great potential for generating electricity by this means but there are many technological problems to be overcome. One advantage of wave power, however, is that it will never run out.

WEATHERING

Weathering is the action of wind, rain, water, ice, frost and chemicals on the surface of a rock. These substances cause the rock to be eroded and the surface of the rock breaks up into smaller particles. These particles are often carried by wind and rivers and deposited in other places.

◀ Sedimentary rocks ▶

WEIGHT

Weight is a result of the gravitational pull of Earth on a mass. Weight is measured in **newtons**, symbol N. The weight of the object varies according to where it is measured. For example, an object would weigh less on the **Moon**, although its mass would stay the same. The **mass** of the Moon is smaller than that of **Earth**, so the gravitational pull of the Moon is about one sixth that of Earth. An object of mass of 10 kg would weigh 16 N on the Moon, whereas on Earth it would weigh 100 N (Fig W.4).

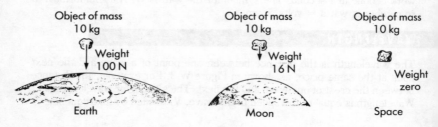

Fig W.4 Weight: the mass of the object stays the same but the weight changes

WHITE BLOOD CELLS

There are about 20 million white blood cells in your body, that's about 7000 for every 1 mm³ of blood. The function of white blood cells is to protect the body against disease, either by producing antibodies which kill bacteria, or by engulfing and destroying bacteria (Fig W.5).

Fig W.5 White blood cells: white blood cells engulf and destroy bacteria in your body

WIND ENERGY

Wind turbines (windmills) convert **kinetic energy** into electricity. Wind turbines are often used on isolated islands where it is difficult to transmit electricity from the mainland. A common wind turbine used in this way may have a very large blade about 20 metres long, and be capable of **generating** enough electricity for about 2000 houses (Fig W.6).

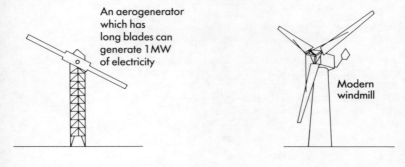

Fig W.6 Wind energy: using wind energy to generate electricity

WORK

Work is done when a **force** moves an object. Work is measured by multiplying the force used by the distance moved, $W = F \times D$. The unit of work is a **joule** (J).

For example, if a person pushes a box with a force of 200 N over a distance of 10 m they have done $200 \times 10 = 2000$ J of work (Fig W.7).

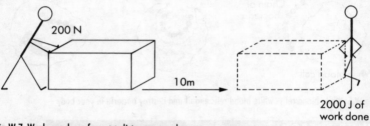

Fig W.7 Work: work = force × distance moved

XENON

Xenon is an inert gas and is in the same group of the **periodic table** (Group 0) as helium, neon and argon. Xenon gas consists of xenon atoms – chemical symbol Xe. It is not very reactive, but will react with fluorine to form a few compounds.

X-RAYS

X-rays are **electromagnetic** waves which have a very short **wavelength**, between 10^{-9} and 10^{-11} of a metre. X-rays are able to pass through many materials and are therefore used to 'see' through dense objects. For example, at many airports suitcases are passed through X-rays to search for any metal objects which may be dangerous. In hospitals, X-rays are used to identify where bones may be broken. You may have had an X-ray photograph taken of your teeth at the dentist.

The X-rays pass through the object and fall onto a photographic plate. The amount that passes through an object depends on the object's density, so the photographic plate shows images of features such as bone etc., but not muscle. X-rays are also used by scientists to help find out the internal structure of materials such as crystals, and used in industry to show hidden flaws in sheet metal.

YEAR

A year is the time taken for the **Earth** to orbit the **Sun**, about 365 days (Fig Y.1). In practice, the orbit takes 365.25 days, so every four years an extra day is added on to the year to compensate for the additional day. The year with the extra day is called a *leap* year and the extra day is February 29th.

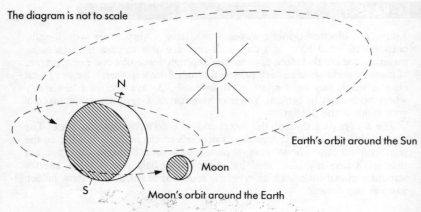

Fig Y.1 Year: how the Earth orbits the Sun

YEAST

Yeast is a single-celled fungus which produces **enzymes**. The enzymes can be used to break down starch and sugar into alcohol and carbon dioxide in the process known as **fermentation**.

During fermentation, yeast cells reproduce by *budding*; small portions of a parent yeast cell can separate to form 'daughter' cells (Fig Y.2). The cells bud to form long chains.

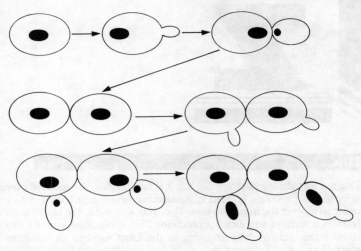

Fig Y.2 Yeast: reproduction by cells budding

ZINC

Zinc is a metallic element symbol Zn. It is fairly high in the **reactivity series** so is quite reactive. Zinc is used in *galvanizing* where it is coated onto iron in order to protect the iron from corrosion. It is also welded in blocks onto ships hulls as a form of **sacrificial protection**. Zinc is obtained from its ore (zinc blende; zinc sulphide) by smelting in the **blast furnace** or sometimes by **electrolysis.** ◄ Corrosion ►

ZYGOTE

A zygote is formed when a female sex cell is **fertilized** by a male sex cell. For example in humans, an egg (ovum) joins with a sperm. The zygote will develop by dividing rapidly and form an **embryo**.